Basic Geometry
for College Students

Basic Geometry for College Students

An Overview of the Fundamental Concepts

SECOND EDITION

Alan S. Tussy
Citrus College

R. David Gustafson
Rock Valley College

BROOKS/COLE
CENGAGE Learning

Australia • Brazil • Japan • Korea • Mexico • Singapore • Spain • United Kingdom • United States

Basic Geometry for College Students:
An Overview of Fundamental Concepts,
Second Edition
Alan S. Tussy and R. David Gustafson

Publisher: Charles Van Wagner

Acquisitions Editor: Marc Bove

Developmental Editor: Danielle Derbenti

Assistant Editor: Shaun Williams

Editorial Assistant: Mary de la Cruz

Media Editor: Sam Subity

Marketing Manager: Greta Kleinhart

Marketing Assistant: Angela Kim

Marketing Communications Manager:
Katherine Malatesta

Content Project Manager: Michelle Cole

Creative Director: Rob Hugel

Art Director: Vernon Boes

Print Buyer: Karen Hunt

Rights Acquisitions Account Manager, Text:
Margaret Chamberlain-Gaston

Rights Acquisitions Account Manager, Image:
Don Schlotman

Production Service: Macmillan Publishing
Solutions

Photo Researcher: Jennifer Lim

Copy Editor: Jill Pellarin

Cover Designer: Teri Wright

Cover Image: ©Wolfgang Volz/laif/Redux

Compositor: Macmillan Publishing Solutions

For product information and technology assistance, contact us at
Cengage Learning Customer & Sales Support, 1-800-354-9706.
For permission to use material from this text or product,
submit all requests online at **www.cengage.com/permissions.**
Further permissions questions can be e-mailed to
permissionrequest@cengage.com.

Library of Congress Control Number: 2008944269

ISBN-13: 978-0-495-82948-5

ISBN-10: 0-495-82948-X

Brooks/Cole
10 Davis Drive
Belmont, CA 94002-3098
USA

Cengage Learning is a leading provider of customized learning solutions with office locations around the globe, including Singapore, the United Kingdom, Australia, Mexico, Brazil, and Japan. Locate your local office at **www.cengage.com/international**.

Cengage Learning products are represented in Canada by Nelson Education, Ltd.

To learn more about Brooks/Cole, visit **www.cengage.com/brookscole**

Purchase any of our products at your local college store or at our preferred online store **www.cengagebrain.com**.

Printed in China by China Translation & Printing Services Limited
2 3 4 5 6 7 13 12 11 10

CONTENTS

For the Instructor

Geometry may be the area of study that is the most overlooked in today's college mathematics curriculum. This is unfortunate; the study of geometry is very beneficial. It develops students' logic and reasoning skills and improves their ability to spot valid and invalid arguments. When studying geometry, students visualize in two and three dimensions, and they gain a better understanding of measurement and units. Most important, students have an opportunity to couple their algebra skills with the geometry concepts that they study to solve meaningful application problems.

Basic Geometry for College Students has been written to address the need for a concise overview of fundamental geometry topics. Sections 1–7, the first part of the text, introduce such topics as angles, polygons, perimeter, area, and circles. In the second part of the text, Sections 8–11 cover congruent and similar triangles, special triangles, volume, and surface area. Appendix I explains inductive and deductive reasoning.

This text can be used for a short introductory course (5–6 weeks) in geometry. It can also serve as a geometry supplement for an elementary algebra course, an intermediate algebra course, or a combination elementary/intermediate algebra course.

Basic Geometry for College Students uses a variety of instructional approaches that reflect the recommendations of NCTM and AMATYC. You will find the vocabulary, practice, and well-defined pedagogy of a traditional approach. We also emphasize the reasoning, modeling, communicating, and technological skills that are such a big part of the current reform movement.

New to this edition

- **Useful Objectives That Help Keep Students Focused**
  Objectives are now numbered at the start of each section to focus students' attention on the skills that they will learn as they work through the section. When each objective is introduced, the number and heading will appear again to remind them of the objective at hand.

- **Examples That Tell Students Not Just How, But WHY**
  Why? Students often ask that question as they watch their instructor solve problems in class and as they are working on problems at home. It's not enough to know how a problem is solved. Students gain a deeper understanding of the algebraic concepts if they know why a particular approach was taken. This instructional truth was the motivation for adding a *Strategy* and *Why* explanation to each worked example.

- **Examples That Ask Students to Try**
  Each example ends with a *Now Try* problem. These are the final step in the learning process. Each one is linked to similar problems found within the *Guided Practice* section of the *Study Sets.*

- **Heavily Revised Study Sets**

  The *Study Sets* have been thoroughly revised to ensure every concept is covered even if the instructor traditionally assigns every other problem. Particular attention was paid to developing a gradual level of progression.

- **Guided Practice**

  All of the problems in the *Guided Practice* portion of the *Study Sets* are linked to an associated worked example from that section. This feature will promote student success by referring them to the proper example(s) if they encounter difficulties solving homework problems.

- **Try It Yourself**

  To promote problem recognition, some *Study Sets* now include a collection of *Try It Yourself* problems that do not have the example linking. The problem types are thoroughly mixed and are not linked, giving students an opportunity to practice decision making and strategy selection as they would when taking a test or quiz.

Acknowledgments

The authors wish to express their gratitude to the Mathematics Department of Century College, White Bear Lake, Minnesota, whose innovative curriculum revision provided the impetus for this project. We offer special thanks to Professor Beth Hentges and Professor Carol Purcell for their helpful input and diligent proofreading. Without the talents and dedication of the editorial, production, and sales staff of Brooks/Cole, this text would not have been so well accomplished. We express our sincere appreciation of the hard work of Charlie Van Wagner, Shaun Williams, Danielle Derbenti, Steve Odrich, and Paul McCombs as well as the freelance talents of David Hoyt and Lori Heckelman and the superb typesetting of Macmillan Publishing Solutions.

Alan S. Tussy
R. David Gustafson

THE WORD GEOMETRY COMES FROM THE GREEK WORDS GEO (MEANING EARTH) AND METRON (MEANING MEASURE).

---

1 *Basic Geometric Figures*

Objectives

1. Identify and name points, lines, and planes.
2. Identify and name line segments and rays.
3. Identify and name angles.
4. Use a protractor to measure angles.

Geometry is a branch of mathematics that studies the properties of two- and three-dimensional figures such as triangles, circles, cylinders, and spheres. More than 5,000 years ago, Egyptian surveyors used geometry to measure areas of land in the flooded plains of the Nile River after heavy spring rains. Even today, engineers marvel at the Egyptians' use of geometry in the design and construction of the pyramids. History records many other practical applications of geometry made by Babylonians, Chinese, Indians, and Romans.

The classical Greeks (660–300 BCE) are credited with refining years of practical use of geometry into a systematic subject of logical thought. **Pythagoras** (580?–501? BCE) is generally regarded as the first of the great Greek mathematicians. Although very little is known about him personally, there is much fascinating literature about the Order of Pythagoreans—a private academic society that he established. Pythagoras and his students devoted themselves to the study of mathematics, astronomy, and philosophy, and they developed geometry into an abstract science.

Many scholars consider **Euclid** (330?–275? BCE) to be the greatest of the Greek mathematicians. His book *The Elements* is an impressive study of geometry and number theory. It presents geometry in a highly structured form that begins with several simple assumptions and then expands on them using logical reasoning. For more than 2,000 years, *The Elements* was the textbook that students all over the world used to learn geometry.

EUCLID

1. Identify and name points, lines, and planes.

Geometry is based on three undefined words: *point, line,* and *plane*. Although we will make no attempt to define these words formally, we can think of a **point** as a geometric figure that has position but no length, width, or depth. Points can be represented on paper by drawing small dots and they are labeled with capital letters. For example, point *A* is shown in figure (a) below.

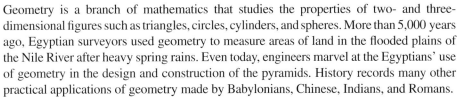

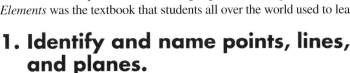

Line *BC* is denoted as $\overleftrightarrow{BC}$.

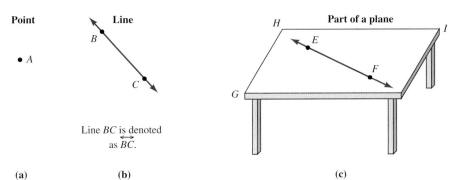

(a)   (b)   (c)

Lines are made up of points. A line extends infinitely far in both directions, but has no width or depth. Lines can be represented on paper by drawing a straight line with arrowheads at either end. In figure (b) on the previous page, the line that passes through points B and C is written as $\overleftrightarrow{BC}$.

Planes are also made up of points. A plane is a flat surface, extending infinitely far in every direction, that has length and width but no depth. The top of a table is a representation of part of a plane. In figure (c) on the previous page, $\overleftrightarrow{EF}$ lies in plane GHI.

As figure (b) illustrates, points B and C determine exactly one line, the line $\overleftrightarrow{BC}$. In figure (c), the points E and F determine exactly one line, the line $\overleftrightarrow{EF}$. In general, any two different points determine exactly one line.

As figure (c) illustrates, points G, H, and I determine exactly one plane. In general, any three different points determine exactly one plane.

Other geometric figures can be created by using parts or combinations of points, lines, and planes.

2. Identify and name line segments and rays.

Line segment

> The **line segment** AB, written as $\overline{AB}$, is the part of a line that consists of points A and B and all points in between (see the figure below). Points A and B are the **endpoints** of the segment.

Line segment AB
is denoted
as $\overline{AB}$.

Every line segment has a **midpoint,** which divides the segment into two parts of equal length. In the figure below, M is the midpoint of segment AB, because the measure of $\overline{AM}$ (written as m($\overline{AM}$)) is equal to the measure of $\overline{MB}$ (written as m($\overline{MB}$)).

$$m(\overline{AM}) = 4 - 1$$
$$= 3$$

and

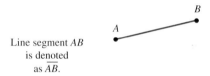

$$m(\overline{MB}) = 7 - 4$$
$$= 3$$

Since the measure of both segments is 3 units, m($\overline{AM}$) = m($\overline{MB}$).

When two line segments have the same measure, we say that they are **congruent.** Since m($\overline{AM}$) = m($\overline{MB}$), we can write

$$\overline{AM} \cong \overline{MB} \quad \text{Read the symbol } \cong \text{ as "is congruent to."}$$

Another geometric figure is the ray, as shown below.

Ray

> A **ray** is the part of a line that begins at some point (say, A) and continues forever in one direction. Point A is the **endpoint** of the ray.

Ray AB is denoted as $\overrightarrow{AB}$. The endpoint
of the ray is always listed first.

To name a ray, we list its endpoint and then one other point on the ray. Sometimes it is possible to name a ray in more than one way. For example, in the figure below, $\overrightarrow{DE}$ and $\overrightarrow{DF}$ name the same ray. This is because both have point D as their endpoint and extend forever in the same direction. In contrast, $\overrightarrow{DE}$ and $\overrightarrow{ED}$ are not the same ray. They have different endpoints and point in opposite directions.

3. Identify and name angles.

Angle

> An **angle** is a figure formed by two rays with a common endpoint. The common endpoint is called the **vertex,** and the rays are called **sides.**

The angle shown below can be denoted as

$\angle BAC$, $\angle CAB$, $\angle A$,   or $\angle 1$ The symbol $\angle$ means angle.

CAUTION   When using three letters to name an angle, be sure the letter name of the vertex is the middle letter. Furthermore, we can only name an angle using a single vertex letter when there is no possibility of confusion. For example, in the figure below, we cannot refer to any of the angles as simply $\angle X$, because we would not know if that meant $\angle WXY$, $\angle WXZ$, or $\angle YXZ$.

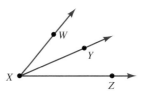

4. Use a protractor to measure angles.

One unit of measurement of an angle is the **degree.** The symbol for degree is a small raised circle, $^\circ$. An angle measure of 1° (read as "one degree") means that one side of an angle is rotated $\frac{1}{360}$ of a complete revolution about the vertex from the other side of the angle. The measure of $\angle ABC$, shown below, is 1°. We can write this in symbols as $m(\angle ABC) = 1^\circ$.

This side of the angle is rotated $\frac{1}{360}$ of a complete revolution from the other side of the angle.

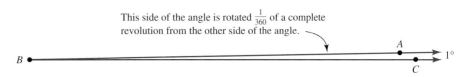

The following figures show the measures of several other angles. An angle measure of 90° is equivalent to $\frac{90}{360} = \frac{1}{4}$ of a complete revolution. An angle measure of 180°

is equivalent to $\frac{180}{360} = \frac{1}{2}$ of a complete revolution, and an angle measure of 270° is equivalent to $\frac{270}{360} = \frac{3}{4}$ of a complete revolution.

$$m(\angle FED) = 90° \qquad m(\angle IHG) = 180° \qquad m(\angle JKL) = 270°$$

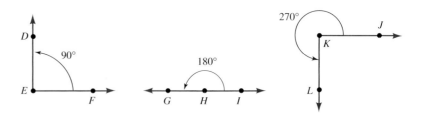

We can use a **protractor** to measure angles. To begin, we place the center of the protractor at the vertex of the angle, with the edge of the protractor aligned with one side of the angle, as shown below. The angle measure is found by determining where the other side of the angle crosses the scale. Be careful to use the appropriate scale, inner or outer, when reading an angle measure.

Angle	Measure in degrees
∠ABC	30°
∠ABD	60°
∠ABE	110°
∠ABF	150°
∠ABG	180°

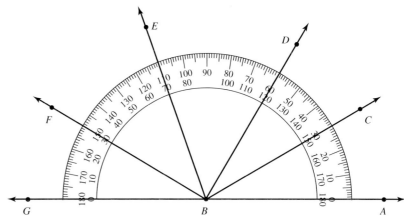

If we read the protractor from left to right, using the inner scale, we can see that m(∠GBF) = 30°. If we read the protractor from right to left, using the outer scale, we see that m(∠ABC) = 30°.

When two angles have the same measure, we say that they are **congruent.** Since m(∠ABC) = 30° and m(∠GBF) = 30°, we can write

$$\angle ABC \cong \angle GBF \qquad \text{Read the symbol } \cong \text{ as "is congruent to."}$$

We classify angles according to their measure.

Classification of angles

Acute angles: Angles whose measures are greater than 0° but less than 90°.

Right angles: Angles whose measures are 90°.

Obtuse angles: Angles whose measures are greater than 90° but less than 180°.

Straight angles: Angles whose measures are 180°.

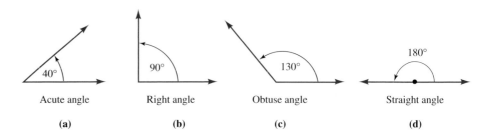

Acute angle	Right angle	Obtuse angle	Straight angle
(a)	**(b)**	**(c)**	**(d)**

A ⌐ symbol is often used in diagrams to denote a right angle. For example, in the figure below, the ⌐ symbol drawn near the vertex of ∠ABC indicates that m(∠ABC) = 90°.

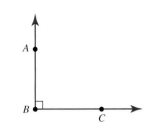

EXAMPLE 1   Classify each angle in the figure as an acute angle, a right angle, an obtuse angle, or a straight angle.

Strategy
We will determine how each angle's measure compares to 90° or to 180°.

Why
Acute, right, obtuse, and straight angles are defined with respect to 90° and 180° angle measures.

Solution
Since m(∠1) < 90°, it is an acute angle.

Since m(∠2) > 90° but less than 180°, it is an obtuse angle.

Since m(∠BDE) = 90°, it is a right angle.

Since m(∠ABC) = 180°, it is a straight angle.

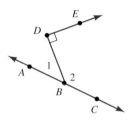

Self Check 1
Classify ∠EFG, ∠DEF, ∠1, and ∠GED in the figure as an acute angle, a right angle, an obtuse angle, or a straight angle.

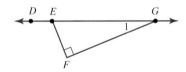

Answer:
right angle, obtuse angle, acute angle, straight angle

Now Try
Problems 41, 43, and 45   ■

STUDY SET Section 1

VOCABULARY *Fill in the blanks.*

1. Three undefined words in geometry are _____, _____, and _____.

2. A line _____ has two endpoints.

3. A _____ divides a line segment into two parts of equal length.

4. A _____ is the part of a line that begins at some point and continues forever in one direction.

5. An _____ is formed by two rays with a common endpoint.

6. An angle is measured in _____ .

7. A _____ is used to measure angles.

8. The measure of an _____ angle is less than 90°.

9. The measure of a _____ angle is 90°.

10. The measure of an _____ angle is greater than 90° but less than 180°.

11. The measure of a straight angle is _____.

12. When two segments have the same length, we say that they are _____. When two angles have the same measure, we say that they are _____.

CONCEPTS

13. a. Given two points (say, *M* and *N*), how many different lines pass through these two points?

   b. Fill in the blank: In general, two different points determine exactly one _____.

14. Refer to the figure.

   a. Name $\overrightarrow{NM}$ in another way.

   b. Do $\overrightarrow{MN}$ and $\overrightarrow{NM}$ name the same ray?

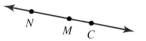

15. Consider the acute angle shown below.

  a. What two rays are the sides of the angle?

  b. What point is the vertex of the angle?

  c. Name the angle in four ways.

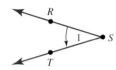

16. Estimate the measure of each angle. Do not use a protractor.

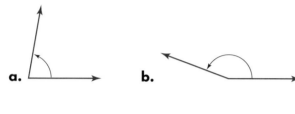

a.   **b.**

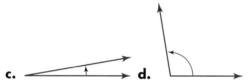

c.   **d.**

17. Draw an example of each type of angle.

  a. an acute angle   **b.** an obtuse angle

  c. a right angle   **d.** a straight angle

18. Fill in the blanks with the correct symbol.

  a. If m($\overline{AB}$) − m($\overline{CD}$), then $\overline{AB}$ __ $\overline{CD}$.

  b. If $\angle ABC \cong \angle DEF$, then m($\angle ABC$) __ m($\angle DEF$).

NOTATION   *Fill in the blanks.*

19. The symbol $\overleftrightarrow{AB}$ is read as "_____ AB."

20. The symbol $\overline{AB}$ is read as "_____ AB."

21. The symbol $\overrightarrow{AB}$ is read as "_____ AB."

22. We read m($\overline{AB}$) as "the _____ of segment AB."

23. We read $\angle ABC$ as "_____ ABC."

24. We read m($\angle ABC$) as "the _____ of angle ABC."

25. The symbol for _____ is a small raised circle, °.

26. The symbol ⌐ indicates a _____ angle.

GUIDED PRACTICE

27. Draw each geometric figure and label it completely. **See Objective 1.**

  a. Point T

  b. $\overleftrightarrow{JK}$

  c. Plane ABC

28. Draw each geometric figure and label it completely. **See Objectives 2 and 3.**

  a. $\overline{RS}$ **b.** $\overrightarrow{PQ}$

  c. $\angle XYZ$ **d.** $\angle L$

Refer to the figure and find the length of each segment. **See Objective 2.**

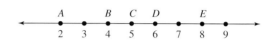

29. $\overline{AB}$ **30.** $\overline{BE}$
31. $\overline{CE}$ **32.** $\overline{BD}$
33. $\overline{DC}$ **34.** $\overline{DE}$
35. $\overline{EA}$ **36.** $\overline{CA}$

Refer to the figure above and find each midpoint. **See Objective 2.**

37. Find the midpoint of $\overline{AD}$.

38. Find the midpoint of $\overline{BD}$.

39. Find the midpoint of $\overline{BE}$.

40. Find the midpoint of $\overline{EA}$.

Classify each angle in the figure as an acute angle, a right angle, an obtuse angle, or a straight angle. **See Example 1.**

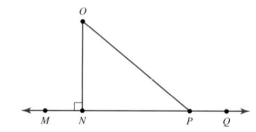

41. $\angle MNO$ **42.** $\angle OPN$
43. $\angle NOP$ **44.** $\angle QPO$
45. $\angle MPQ$ **46.** $\angle ONM$
47. $\angle QPO$ **48.** $\angle QPN$

Use the protractor to find each angle measure listed on the next page. **See Objective 4.**

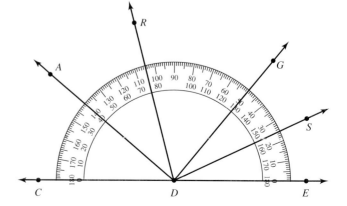

49. m(∠GDE)   **50.** m(∠ADE)

51. m(∠EDS)   **52.** m(∠EDR)

53. m(∠CDR)   **54.** m(∠CDA)

55. m(∠CDG)   **56.** m(∠CDS)

57. m(∠CDE)   **58.** m(∠EDC)

59. m(∠ADG)   **60.** m(∠GDR)

TRY IT YOURSELF   *Refer to the figure below and tell whether each statement is true. If a statement is false, explain why.*

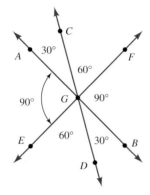

61. $\overrightarrow{GF}$ has point *G* as its endpoint.

62. $\overline{AG}$ has no endpoints.

63. $\overleftrightarrow{CD}$ has three endpoints.

64. Point *D* is the vertex of ∠DGB.

65. m(∠AGC) = m(∠BGD)

66. ∠AGF ≅ ∠BGE

67. ∠FGB ≅ ∠EGA

68. ∠AGC and ∠FGC share a common side.

Refer to the figure above and tell whether each angle is an acute angle, a right angle, an obtuse angle, or a straight angle.

69. ∠AGC   **70.** ∠EGA

71. ∠FGD   **72.** ∠BGA

73. ∠BGE   **74.** ∠AGD

75. ∠DGC   **76.** ∠DGB

Use a protractor to measure each angle. **See Objective 4.**

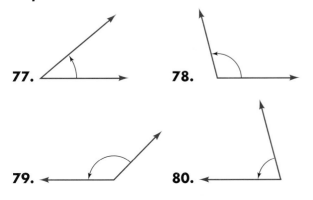

77.   **78.**

79.   **80.**

81.   **82.**

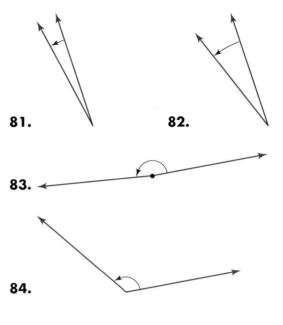

83.

84.

APPLICATIONS

85. BASEBALL   Use the following official definition to draw the strike zone for the player shown below: The *strike zone* is that area over home plate the upper limit of which is a horizontal line at the midpoint between the top of the shoulders and the top of the uniform pants, and the lower limit is a line at the hollow beneath the kneecap.

86. PHYSICS   The illustration below shows a 15-pound block that is suspended with two ropes, one of which is horizontal. Classify each numbered angle as acute, obtuse, or right.

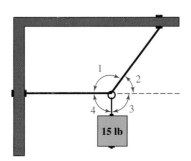

15 lb

87. MUSICAL INSTRUMENTS   Suppose that you are a beginning band teacher describing the correct posture needed to play various instruments. Using the diagrams shown below, approximate the angle measure (in degrees) at which each instrument should be held in relation to the student's body.

a. flute   **b.** clarinet   **c.** trumpet

a.   b.   c.

88. PLANETS   The figures below show the direction of rotation of several planets in our solar system. They also show the angle of tilt of each planet.

a. Which planets have an angle of tilt that is an acute angle?

b. Which planets have an angle of tilt that is an obtuse angle?

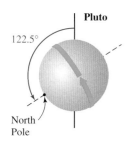 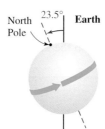

 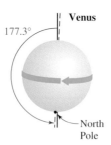

WRITING

89. PHRASES   Explain what you think each of these phrases means. How is geometry involved?

a. The president did a complete 180-degree flip on the subject of a tax cut.

b. The rollerblader did a "360" as she jumped off the ramp.

90. In the statements below, the ° symbol is used in two different ways. Explain the difference.

$$m(\angle A) = 85° \qquad \text{and} \qquad 85°F$$

91. What is a protractor?

92. Explain the difference between a ray and a line segment.

93. Explain why we *cannot* refer to any of the angles in the figure below as $\angle T$.

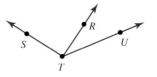

94. Describe what is wrong with the drawing of $\angle R$ shown below.

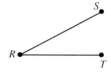

2 *More about Angles*

Objectives

1. Solve problems involving adjacent angles.
2. Use the property of vertical angles to solve problems.
3. Solve problems involving complementary and supplementary angles.

The drawings below show three ways in which rays and lines can intersect to form angles. Often, a pair (or pairs) of angles in drawings like these are related. In this section, we will explore some important relationships that exist between certain pairs of angles.

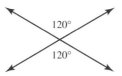
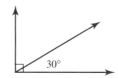
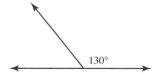

1. Solve problems involving adjacent angles.

Two angles that have a common vertex and a common side are called **adjacent angles** if they are side-by-side and their interiors do not overlap.

EXAMPLE 1 Two angles with degree measures of x and $35°$ are adjacent angles, as shown. Use the information in the figure to find x.

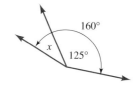

Strategy
We will write an equation involving x that mathematically models the situation.

Why
We can then solve the equation to find the unknown angle measure.

Solution
Since the sum of the measures of the two adjacent angles is $80°$, we have

$$x + 35° = 80°$$ The word *sum* indicates addition.

$$x + 35° - 35° = 80° - 35°$$ To isolate x, undo the addition of $35°$ by subtracting $35°$ from both sides.

$$x = 45°$$ Do the subtractions: $35° - 35° = 0°$ and $80° - 35° = 45°$.

Thus, x is $45°$. As a check, we see that $45° + 35° = 80°$.

Self Check 1
Use the information in the figure to find x.

Answer: $35°$

Now Try
Problem 33

In the figure for Example 1, we used the variable x to represent an unknown angle measure. In such cases, we will assume that the variable "carries" with it the associated units of degrees. That means we do not have to write a ° symbol next to the variable. Furthermore, if x represents an unknown number of degrees, then expressions such as $3x$, $x + 15°$, and $4x - 20°$ also have units of degrees.

2. Use the property of vertical angles to solve problems.

When two lines intersect, pairs of nonadjacent angles are called **vertical angles.** In the figure below, lines l_1 (read as "line l sub 1") and l_2 (read as "line l sub 2") intersect. $\angle 1$ and $\angle 3$ are vertical angles, as are $\angle 2$ and $\angle 4$.

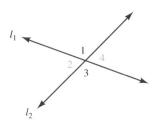

To illustrate that vertical angles always have the same measure, refer to the following figure, with angles having measures of x, y, and 30°. Since the measure of any straight angle is 180°, we have

$$30° + x = 180° \qquad \text{and} \qquad 30° + y = 180°$$

$$x = 150° \qquad\qquad\qquad y = 150° \quad \text{To undo the addition of 30°, subtract}$$
$$\text{30° from both sides.}$$

Since x and y are both 150°, we conclude that $x = y$.

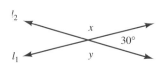

Note that the angles having measures x and y are vertical angles.

The previous example illustrates that vertical angles have the same measure. Recall that when two angles have the same measure, we say that they are *congruent*. Therefore, we have the following important fact:

Property of vertical angles	Vertical angles are congruent (have the same measure).

EXAMPLE 2   Refer to the figure. Find
a. m($\angle 1$) and **b.** m($\angle ABF$).

Strategy
To answer part a, we will use the property of vertical angles. To answer part b, we will write an equation involving m($\angle ABF$) that mathematically models the situation.

Why
For part a, we note that $\overleftrightarrow{AD}$ and $\overleftrightarrow{BC}$ intersect to form vertical angles. For part b, we can solve the equation to find the unknown, m($\angle ABF$).

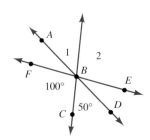

Self Check 2
Refer to the figure for Example 2. Find

a. m($\angle 2$)

b. m($\angle DBE$)

Solution

a. If we ignore $\overleftrightarrow{FE}$ for the moment, we see that $\overleftrightarrow{AD}$ and $\overleftrightarrow{BC}$ intersect to form the pair of vertical angles $\angle CBD$ and $\angle 1$. By the property of vertical angles,

$$\angle CBD \cong \angle 1 \qquad \text{Read as "Angle } CBD \text{ is congruent to angle one."}$$

Since congruent angles have the same measure,

$$\mathrm{m}(\angle CBD) = \mathrm{m}(\angle 1)$$

In the figure, we are given $\mathrm{m}(\angle CBD) = 50°$. Thus, $\mathrm{m}(\angle 1)$ is also $50°$, and we can write $\mathrm{m}(\angle 1) = 50°$.

b. Since $\angle ABD$ is a straight angle, the *sum* of the measures of $\angle ABF$, the $100°$ angle, and the $50°$ angle is $180°$. If we let $x = \mathrm{m}(\angle ABF)$, we have

$$
\begin{array}{ll}
x + 100° + 50° = 180° & \text{The word } \textit{sum} \text{ indicates addition.} \\
x + 150° = 180° & \text{On the left side, combine like terms:} \\
& 100° + 50° = 150°. \\
x = 30° & \text{To isolate } x, \text{ undo the addition of } 150° \text{ by subtracting} \\
& 150° \text{ from both sides: } 180° - 150° = 30°.
\end{array}
$$

Thus, $\mathrm{m}(\angle ABF) = 30°$

EXAMPLE 3   In the figure on the right, find **a.** x, **b.** $\mathrm{m}(\angle ABC)$, and **c.** $\mathrm{m}(\angle CBE)$.

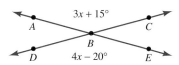

Strategy

We will use the property of vertical angles to write an equation that mathematically models the situation.

Why

$\overleftrightarrow{AE}$ and $\overleftrightarrow{DC}$ intersect to form two pairs of vertical angles.

Solution

a. In the figure, two vertical angles have degree measures that are represented by the algebraic expressions $4x - 20°$ and $3x + 15°$. Since the angles are vertical angles, they have equal measures.

$$
\begin{array}{ll}
4x - 20° = 3x + 15° & \text{Set the algebraic expressions equal.} \\
4x - 20° - 3x = 3x + 15° - 3x & \text{To eliminate } 3x \text{ from the right side,} \\
& \text{subtract } 3x \text{ from both sides.} \\
x - 20° = 15° & \text{Combine like terms: } 4x - 3x = x \text{ and} \\
& 3x - 3x = 0. \\
x = 35° & \text{To isolate } x, \text{ undo the subtraction of } 20° \\
& \text{by adding } 20° \text{ to both sides.}
\end{array}
$$

Thus, x is $35°$.

b. To find $\mathrm{m}(\angle ABC)$, we evaluate the expression $3x + 15°$ for $x = 35°$.

$$
\begin{array}{ll}
3x + 15° = 3(35°) + 15° & \text{Substitute } 35° \text{ for } x. \\
= 105° + 15° & \text{Do the multiplication.} \\
= 120°
\end{array}
$$

Thus, $\mathrm{m}(\angle ABC) = 120°$.

c. $\angle ABE$ is a straight angle. Since the measure of a straight angle is $180°$ and $\mathrm{m}(\angle ABC) = 120°$, $\mathrm{m}(\angle CBE)$ must be $180° - 120°$, or $60°$.

Answers: **a.** $100°$ **b.** $30°$

Now Try

Problems 37 and 39

Self Check 3

In the figure below, find

a. y

b. $\mathrm{m}(\angle XYZ)$

c. $\mathrm{m}(\angle MYX)$

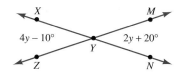

Answers: **a.** $15°$ **b.** $50°$ **c.** $130°$

Now Try

Problem 43

For more complicated problems, it is often helpful to use the following five-step problem-solving strategy.

EXAMPLE 4   Two angles are vertical angles. The measure of the first angle is 5 degrees more than a number, and the measure of the second angle is 25° less than six times the number. What is the measure of the first angle?

Analyze the problem
- There are two angles to consider, and they are vertical angles.
- The measure of one angle is 5° more than a number.
- The measure of the other angle is 25° less than six times the same number.
- We are to find the measure of the first angle.

Form an equation   Since the measure of each angle is expressed in terms of the same number, we begin by letting n = the number. When we translate the key phrases *5° more than a number* and *25° less than six times the number,* we see that each angle measure can be represented by an algebraic expression.

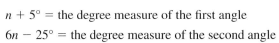

$n + 5°$ = the degree measure of the first angle

$6n - 25°$ = the degree measure of the second angle

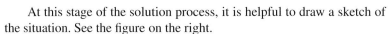

At this stage of the solution process, it is helpful to draw a sketch of the situation. See the figure on the right.

Because the angles are vertical angles, they are congruent, and we have

The measure of the first angle	is equal to	the measure of the second angle.
$n + 5°$	$=$	$6n - 25°$

Solve the equation

$n + 5° = 6n - 25°$	
$n + 5° - n = 6n - 25° - n$	To eliminate n on the left side, subtract n from both sides.
$5° = 5n - 25°$	Combine like terms: $n - n = 0$ and $6n - n = 5n$.
$5° + 25° = 5n - 25° + 25°$	To isolate the variable term, $5n$, undo the subtraction of 25° by adding 25° to both sides.
$30° = 5n$	Do the additions.
$\dfrac{30°}{5} = \dfrac{5n}{5}$	To isolate n, undo the multiplication by 5 by dividing both sides by 5.
$6° = n$	Do the divisions.

The number is 6°. To find the measure of the first angle, we evaluate the expression $n + 5°$ for $n = 6°$.

$n + 5° = 6° + 5°$ Substitute 6° for n.

$= 11°$

State the conclusion   The measure of the first angle is 11°.

Check the result   If we evaluate $6n - 25°$ for $n = 6°$ to find the measure of the other angle, we get $6(6°) - 25° = 11°$. The measures of the vertical angles are the same, 11°. The result checks.

Self Check 4

Two angles are vertical angles. The measure of the first angle is 8° more than a number, and the measure of the second angle is 32° less than twice the number. What is the measure of the first angle?

Answer: 48°

Now Try

Problem 45

3. Solve problems involving complementary and supplementary angles.

Complementary and supplementary angles

> Two angles are **complementary angles** when the sum of their measures is 90°.
> Two angles are **supplementary angles** when the sum of their measures is 180°.

In figure (a) below, ∠*ABC* and ∠*CBD* are complementary angles because the sum of their measures is 90°. Each angle is said to be the **complement** of the other. In figure (b) below, ∠*X* and ∠*Y* are also complementary angles, because m(∠*X*) + m(∠*Y*) = 90°. Figure (b) illustrates an important fact: Complementary angles need not be adjacent angles.

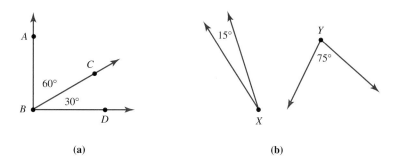

(a)   (b)

In figure (a) below, ∠*MNO* and ∠*ONP* are supplementary angles, because the sum of their measures is 180°. Each angle is said to be the **supplement** of the other. Supplementary angles need not be adjacent angles. For example, in figure (b) below, ∠*G* and ∠*H* are supplementary angles, because m(∠*G*) + m(∠*H*) = 180°.

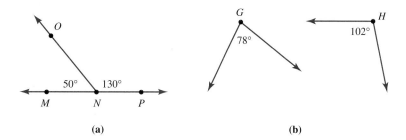

(a)   (b)

 CAUTION The definition of supplementary angles requires that the sum of *two* angles be 180°. Three angles of 40°, 60°, and 80° are not supplementary even though their sum is 180°.

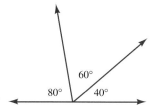

EXAMPLE 5

a. Find the complement of a 35° angle.

b. Find the supplement of a 105° angle.

Strategy

We will use the definitions of complementary and supplementary angles to write equations that mathematically model each situation.

Self Check 5

a. Find the complement of a 50° angle.

b. Find the supplement of a 50° angle.

Why

We can then solve each equation to find the unknown angle measure.

Solution

a. It is helpful to draw a figure, as shown to the right. Let x represent the measure of the complement of the 35° angle. Since the angles are complementary, we have

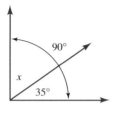

$$x + 35° = 90°$$ The sum of the angles' measures must be 90°.

$$x = 55°$$ To isolate x, undo the addition of 35° by subtracting 35° from both sides: $90° - 35° = 55°$.

The complement of a 35° angle has measure 55°.

b. It is helpful to draw a figure, as shown on the right. Let y represent the measure of the supplement of the 105° angle. Since the angles are supplementary, we have

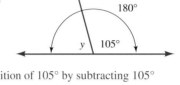

$$y + 105° = 180°$$ The sum of the angles' measures must be 180°.

$$y = 75°$$ To isolate y, undo the addition of 105° by subtracting 105° from both sides: $180° - 105° = 75°$.

The supplement of a 105° angle has measure 75°.

Answers: **a.** 40°   **b.** 130°

Now Try

Problems 49 and 51

EXAMPLE 6   The measure of an angle is 15° more than twice the measure of its supplement. What is the measure of the angle?

Analyze the problem
- There are two angles to consider.
- The angles are supplementary.
- The measure of one angle (the first angle) is 15° more than twice the measure of the other angle (its supplement).
- We are to find the measure of the first angle.

Form an equation   Since the measure of one of the angles is expressed in terms of its supplement, we let x = the measure of the supplement. Translating the given key words *more than* and *twice,* we can represent the measure of the first angle by the expression

$$2x + 15°$$

At this stage of the solution process, it is helpful to draw a sketch of the situation. Because the angles are supplements, we have

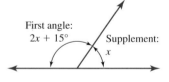

The measure of the first angle	plus	the measure of its supplement	is	180°.
$2x + 15°$	$+$	x	$=$	180°

Solve the equation

$$2x + 15° + x = 180°$$

$$3x + 15° = 180°$$ Combine like terms: $2x + x = 3x$.

$$3x + 15° - 15° = 180° - 15°$$ To isolate the variable term $3x$, undo the addition of 15° by subtracting 15° from both sides.

$$3x = 165°$$ Do the subtractions.

$$\frac{3x}{3} = \frac{165°}{3}$$ To isolate x, undo the multiplication by 3 by dividing both sides by 3.

$$x = 55°$$ Do the divisions.

The measure of the supplement is 55°. To find the measure of the first angle, we evaluate the expression $2x + 15°$ for $x = 55°$.

$$2x + 15° = 2(\mathbf{55°}) + 15° \quad \text{Substitute 55° for } x.$$
$$= 110° + 15° \quad \text{Do the multiplication.}$$
$$= 125°$$

State the conclusion   The measure of the first angle is 125°.

Check the result   Since 125° is 15° more than twice 55°, and since 125° + 55° = 180°, the result 125° checks.

Self Check 6

The measure of an angle is 5° more than six times the measure of its supplement. What is the measure of the angle?

Answer: 155°

Now Try

Problem 53

STUDY SET   Section 2

VOCABULARY   *Fill in the blanks.*

1. _____ angles have the same vertex, are side-by-side, and their interiors do not overlap.

2. When two lines intersect, pairs of nonadjacent angles are called _____ angles.

3. When two angles have the same measure, we say that they are _____ .

4. The word *sum* indicates the operation of _____ .

5. The sum of two complementary angles is _____ .

6. The sum of two _____ angles is 180°.

CONCEPTS

7. a. Draw a pair of adjacent angles. Label them ∠ABC and ∠CBD.

  b. Draw two intersecting lines. Label them lines l_1 and l_2. Label one pair of vertical angles that are formed as ∠1 and ∠2.

  c. Draw two adjacent complementary angles.

  d. Draw two adjacent supplementary angles.

8. Fill in the blanks.

  a. If ∠MNO ≅ ∠BFG, then m(∠MNO) _____ m(∠BFG).

  b. The vertical angle property: Vertical angles are

  _____ .

9. Fill in the blanks.

  a. An angle with measure 33° more than a number n can be represented by the expression:

  $$n +$$

b. An angle with measure 5° less than twice a number n can be represented by the expression:

$$- 5°$$

10. Refer to the figure below. Fill in the blanks.

  a. ∠XYZ and ∠_____ are vertical angles.

  b. ∠XYZ and ∠ZYW are _____ angles.

  c. ∠ZYW and ∠XYV are _____ angles.

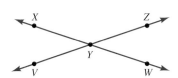

Refer to the figure below and tell whether each statement is true.

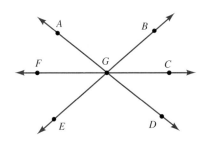

11. ∠AGF and ∠BGC are vertical angles.

12. ∠FGE and ∠BGA are adjacent angles.

13. m($\angle AGB$) = m($\angle BGC$).

14. $\angle AGC \cong \angle DGF$.

Refer to the figure below and tell whether the angles are congruent.

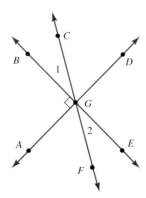

15. $\angle 1$ and $\angle 2$

16. $\angle FGB$ and $\angle CGE$

17. $\angle AGB$ and $\angle DGE$

18. $\angle CGD$ and $\angle CGB$

19. $\angle AGF$ and $\angle FGE$

20. $\angle AGB$ and $\angle BGD$

Refer to the figure above and tell whether each statement is true.

21. $\angle 1$ and $\angle CGD$ are adjacent angles.

22. $\angle 2$ and $\angle 1$ are adjacent angles.

23. $\angle FGA$ and $\angle AGC$ are supplementary.

24. $\angle AGB$ and $\angle BGC$ are complementary.

25. $\angle AGF$ and $\angle 2$ are complementary.

26. $\angle AGB$ and $\angle EGD$ are supplementary.

27. $\angle EGD$ and $\angle DGB$ are supplementary.

28. $\angle DGC$ and $\angle DGE$ are complementary.

NOTATION *Fill in the blanks.*

29. The symbol $\angle$ means _____ .

30. The symbol $\cong$ is read as "is _____ to."

31. A _____ is a letter, such as x, that is used to stand for a number.

32. The symbol l_1 can be used to name a line. It is read as "line l _____ 1."

GUIDED PRACTICE

Find x. See Example 1.

33.

34.

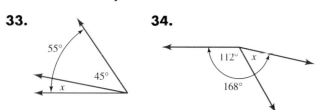

35.

36.

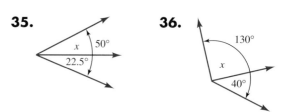

Refer to the figure below. Find the measure of each angle. See Example 2.

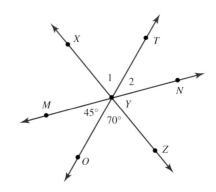

37. $\angle 1$

38. $\angle MYX$

39. $\angle NYZ$

40. $\angle 2$

First find x. Then find m($\angle ABD$) and m($\angle DBE$). See Example 3.

41.

42.

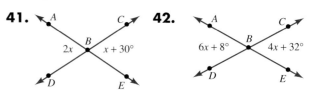

First find x. Then find m($\angle ZYQ$) and m($\angle PYQ$). See Example 3.

43.

44.

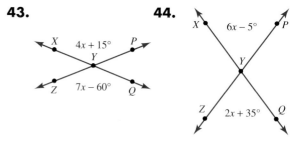

Write an equation to solve each problem. See Example 4.

45. Two angles are vertical angles. The first angle has a measure that is 50° more than a number. The other angle has a measure that is 23° more than ten times the number. Find the measure of the first angle.

46. Two angles are vertical angles. The first angle has a degree measure that is three times a number. The other angle has a measure that is 30° less than five times the number. Find the measure of the first angle.

47. The measure of one angle is 15° more than twice a number, and an angle vertical to it measures 5° less than four times the number. What is the measure of the first angle?

48. The degree measure of one angle is half of a number, and an angle vertical to it measures 15° less than the number. What is the measure of the first angle?

*Let x represent the unknown angle measure. Write an equation and solve it to find x. **See Example 5.***

49. Find the complement of a 30° angle.

50. Find the supplement of a 30° angle.

51. Find the supplement of a 105° angle.

52. Find the complement of a 75° angle.

*Write an equation to solve each problem. **See Example 6.***

53. The measure of an angle is 10° more than the measure of its supplement. What is the measure of the angle?

54. The measure of an angle is 110° more than the measure of its supplement. What is the measure of the angle?

55. The measure of an angle is 8° less than the measure of its complement. What is the measure of the angle?

56. The measure of an angle is 27° less than the measure of its complement. What is the measure of the angle?

TRY IT YOURSELF

Refer to the figure below, in which m(∠1) = 50°. *Find the measure of each angle or sum of angles.*

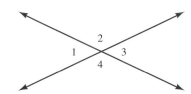

57. ∠3   **58.** ∠4

59. m(∠1) + m(∠2) + m(∠3)

60. m(∠2) + m(∠4)

Refer to the figure below, in which m(∠1) + m(∠3) + m(∠4) = 180°, ∠3 ≅ ∠4, *and* ∠4 ≅ ∠5. *Find the measure of each angle listed in the next column.*

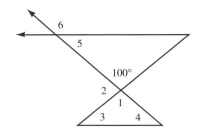

61. ∠1   **62.** ∠2

63. ∠3   **64.** ∠6

Refer to the figure below where ∠1 ≅ ∠ACD, ∠1 ≅ ∠2, *and* ∠BAC ≅ ∠2.

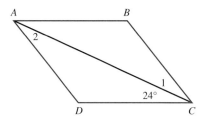

65. What is the complement of ∠BAC?

66. What is the supplement of ∠BAC?

Refer to the figure below where ∠EBS ≅ ∠BES.

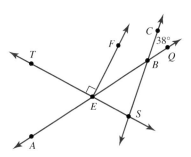

67. What is the measure of ∠AEF?

68. What is the supplement of ∠AET?

69. Find the supplement of the complement of a 51° angle.

70. Find the complement of the supplement of a 173° angle.

71. Find the complement of the complement of a 1° angle.

72. Find the supplement of the supplement of a 6° angle.

73. The measure of an angle is 10° more than three times the measure of its complement. What is the measure of the angle?

74. The measure of an angle is one-half the measure of its supplement. What is the measure of the angle?

75. The measure of an angle is 20° less than nineteen times the measure of its supplement. What is the measure of the angle?

76. The measure of an angle is eight times as large as the measure of its complement. What is the measure of the angle?

APPLICATIONS

77. SYNTHESIZER   Find *x* and *y*.

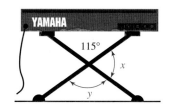

78. AVIATION   How many degrees from the horizontal position are the wings of the airplane?

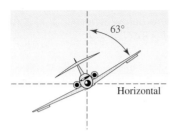

79. GARDENING   What angle does the handle of the lawn mower make with the ground?

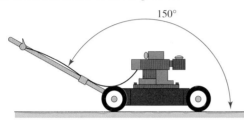

80. ANGLES   Each of the following drawings illustrates a real-life application of a concept that was discussed in this section. Tell what concept is illustrated.

a. Railroad crossing guard

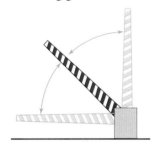

b. Miter saw, used to cut molding

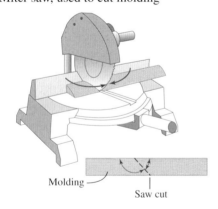

c. Chair

WRITING

81. Explain why an angle measuring 105° cannot have a complement.

82. Explain why an angle measuring 210° cannot have a supplement.

83. Can two angles that are complementary be equal? Explain.

84. Can two angles that are supplementary be equal? Explain.

85. In parts *a* and *b*, explain why the two angles aren't considered to be adjacent angles.

a.   **b.**

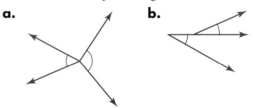

c. Explain why the angles below are not vertical angles.

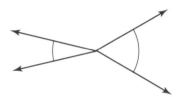

86. Some of the vocabulary introduced in this section can be used in other contexts. In your own words, explain what is meant by each sentence.

a. A homeowner decided to purchase the vacant lot *adjacent* to his property.

b. A teacher decided to *supplement* his income by working a second job on weekends.

c. The right beverage can *complement* a well-cooked meal.

3 *Parallel and Perpendicular Lines*

Objectives

1. Identify and define parallel and perpendicular lines.
2. Identify corresponding angles, interior angles, and alternate interior angles.
3. Use properties of parallel lines cut by a transversal to find unknown angle measures.
4. Use the converse to show that lines are parallel.

In this section, we will consider *parallel* and *perpendicular* lines. Since parallel lines are always the same distance apart, the railroad tracks shown in figure (a) illustrate one application of parallel lines. Figure (b) shows one of the events of men's gymnastics, the parallel bars. Since perpendicular lines meet and form right angles, the monument and the ground shown in figure (c) illustrate one application of perpendicular lines.

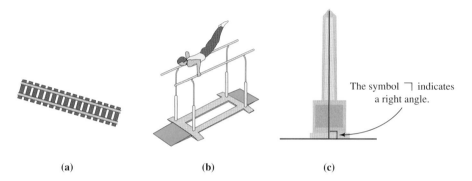

The symbol ⌐ indicates a right angle.

| (a) | (b) | (c) |

1. Identify and define parallel and perpendicular lines.

If two lines lie in the same plane, they are called **coplanar.** Two coplanar lines that do not intersect are called **parallel lines.** See figure (a) below. If two lines do not lie in the same plane, they are called noncoplanar. Two noncoplanar lines that do not intersect are called **skew lines.**

Parallel lines

> Parallel lines are coplanar lines that do not intersect.

If lines l_1 (read as "l sub 1") and l_2 (read as "l sub 2") are parallel, we can write $l_1 \parallel l_2$, where the symbol $\parallel$ is read as "is parallel to."

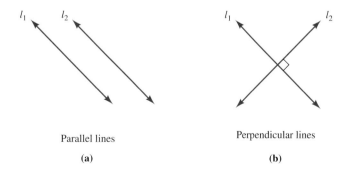

Parallel lines

(a)

Perpendicular lines

(b)

Perpendicular lines

> Perpendicular lines are lines that intersect and form right angles.

In figure (b) above, $l_1 \perp l_2$, where the symbol $\perp$ is read as "is perpendicular to."

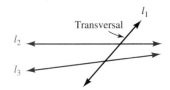

2. Identify corresponding angles, interior angles, and alternate interior angles.

A line that intersects two coplanar lines in two distinct (different) points is called a **transversal.** For example, line l_1 in the figure to the left is a transversal intersecting lines l_2 and l_3.

When two lines are cut by a transversal, all eight angles that are formed are important in the study of parallel lines. Descriptive names are given to several pairs of these angles.

In the figure, four pairs of **corresponding angles** are formed.

Corresponding angles

$\angle 1$ and $\angle 5$

$\angle 3$ and $\angle 7$

$\angle 2$ and $\angle 6$

$\angle 4$ and $\angle 8$

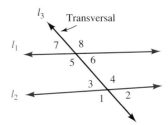

Corresponding angles

> If two lines are cut by a transversal, then the angles on the same side of the transversal and in corresponding positions with respect to the lines are called corresponding angles.

In the figure, four **interior angles** are formed.

Interior angles

$\angle 3$, $\angle 4$, $\angle 5$, and $\angle 6$

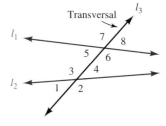

In the figure, two pairs of **alternate interior angles** are formed.

Alternate interior angles

$\angle 4$ and $\angle 5$

$\angle 3$ and $\angle 6$

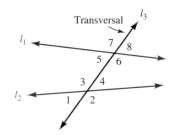

Alternate interior angles

> If two lines are cut by a transversal, then the nonadjacent angles on opposite sides of the transversal and on the interior of the two lines are called alternate interior angles.

 CAUTION Alternate interior angles are easily spotted because they form a Z-shape or a backward Z-shape, as shown below.

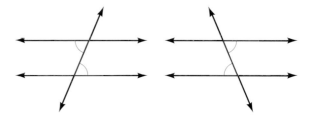

EXAMPLE 1   Refer to the figure. Identify **a.** all pairs of corresponding angles, **b.** all interior angles, and **c.** all pairs of alternate interior angles.

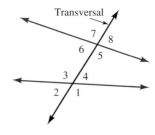

Strategy

When two lines are cut by a transversal, eight angles are formed. We will consider the relative position of the angles with respect to the two lines and the transversal.

Why

There are four pairs of corresponding angles, four interior angles, and two pairs of alternate interior angles.

Solution

a. To identify corresponding angles, we examine the angles to the right of the transversal and the angles to the left of the transversal. The pairs of corresponding angles in the figure are

$\angle 1$ and $\angle 5$, $\angle 4$ and $\angle 8$,

$\angle 2$ and $\angle 6$, $\angle 3$ and $\angle 7$

b. To identify the interior angles, we determine the angles within the two lines cut by the transversal. The interior angles in the figure are

$\angle 3$, $\angle 4$, $\angle 5$, and $\angle 6$

c. Alternate interior angles are angles on opposite sides of the transversal within the two lines. Thus, the pairs of alternate interior angles in the figure are

$\angle 3$ and $\angle 5$, $\angle 4$ and $\angle 6$

Self Check 1

Refer to the figure below. Identify

a. all pairs of corresponding angles,

b. all interior angles

c. all pairs of alternate interior angles.

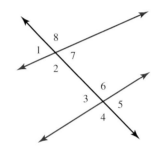

Answers: **a.** $\angle 1$ and $\angle 3$, $\angle 2$ and $\angle 4$, $\angle 8$ and $\angle 6$, $\angle 7$ and $\angle 5$
b. $\angle 2$, $\angle 7$, $\angle 3$, and $\angle 6$
c. $\angle 2$ and $\angle 6$, $\angle 7$ and $\angle 3$

Now Try

Problem 21

3. Use properties of parallel lines cut by a transversal to find unknown angle measures.

Lines that are cut by a transversal may or may not be parallel. When a pair of parallel lines are cut by a transversal, we can make several important observations about the angles that are formed.

1. **Corresponding angles property:** If two parallel lines are cut by a transversal, each pair of corresponding angles are congruent. In the figure below, if $l_1 \parallel l_2$, then $\angle 1 \cong \angle 5$, $\angle 3 \cong \angle 7$, $\angle 2 \cong \angle 6$, and $\angle 4 \cong \angle 8$.

2. **Alternate interior angles property:** If two parallel lines are cut by a transversal, alternate interior angles are congruent. In the figure below, if $l_1 \parallel l_2$, then $\angle 3 \cong \angle 6$ and $\angle 4 \cong \angle 5$.

3. **Interior angles property:** If two parallel lines are cut by a transversal, interior angles on the same side of the transversal are supplementary. In the figure below, if $l_1 \parallel l_2$, then $\angle 3$ is supplementary to $\angle 5$ and $\angle 4$ is supplementary to $\angle 6$.

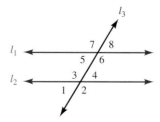

4. If a transversal is perpendicular to one of two parallel lines, it is also perpendicular to the other line. In figure (a) on the next page, if $l_1 \parallel l_2$ and $l_3 \perp l_1$, then $l_3 \perp l_2$.

5. If two lines are parallel to a third line, they are parallel to each other. In figure (b) below, if $l_1 \parallel l_2$ and $l_1 \parallel l_3$, then $l_2 \parallel l_3$.

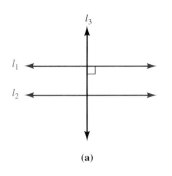

(a)

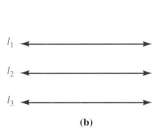

(b)

EXAMPLE 2   Refer to the figure. If $l_1 \parallel l_2$ and $m(\angle 3) = 120°$, find the measures of the other seven angle that are labeled.

Strategy
We will look for vertical angles, supplementary angles, and alternate interior angles in the figure.

Why
The facts that we have studied about vertical angles, supplementary angles, and alternate interior angles enable us to use known angle measures to find unknown angle measures.

Solution

$m(\angle 1) = 60°$	$\angle 3$ and $\angle 1$ are supplementary.
$m(\angle 2) = 120°$	Vertical angles are congruent: $m(\angle 2) = m(\angle 3)$.
$m(\angle 4) = 60°$	Vertical angles are congruent: $m(\angle 4) = m(\angle 1)$.
$m(\angle 5) = 60°$	If two parallel lines are cut by a transversal, alternate interior angles are congruent: $m(\angle 5) = m(\angle 4)$.
$m(\angle 6) = 120°$	If two parallel lines are cut by a transversal, alternate interior angles are congruent: $m(\angle 6) = m(\angle 3)$.
$m(\angle 7) = 120°$	Vertical angles are congruent: $m(\angle 7) = m(\angle 6)$.
$m(\angle 8) = 60°$	Vertical angles are congruent: $m(\angle 8) = m(\angle 5)$.

Some geometric figures contain two transversals.

EXAMPLE 3   Refer to the figure. If $\overline{AB} \parallel \overline{DE}$, which pairs of angles are congruent?

Strategy
We will use the corresponding angles property twice to find two pairs of congruent angles.

Why
Both $\overleftrightarrow{AC}$ and $\overleftrightarrow{BC}$ are transversals cutting the parallel line segments $\overline{AB}$ and $\overline{DE}$.

Solution
Since $\overline{AB} \parallel \overline{DE}$, and $\overleftrightarrow{AC}$ is a transversal cutting them, corresponding angles are congruent. So we have

$$\angle A \cong \angle 1$$

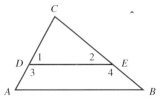

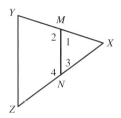

Self Check 2
Refer to the figure for Example 2. If $l_1 \parallel l_2$ and $m(\angle 8) = 50°$, find the measures of the other seven angles that are labeled.

Answer: $m(\angle 5) = 50°$, $m(\angle 7) = 130°$, $m(\angle 6) = 130°$, $m(\angle 3) = 130°$, $m(\angle 4) = 50°$, $m(\angle 1) = 50°$, $m(\angle 2) = 130°$

Now Try
Problem 23

Self Check 3
See the figure below. If $\overline{YZ} \parallel \overline{MN}$, which pairs of angles are congruent?

Since $\overline{AB} \parallel \overline{DE}$ and $\overleftrightarrow{BC}$ is a transversal cutting them, corresponding angles must be congruent. So we have

$$\angle B \cong \angle 2$$

Answer: $\angle 1 \cong \angle Y$, $\angle 3 \cong \angle Z$

Now Try
Problem 25

EXAMPLE 4   In the figure, $l_1 \parallel l_2$. Find x.

Strategy
We will use the corresponding angles property to write an equation that mathematically models the situation.

Why
We can then solve the equation to find x.

Solution
In the figure, two corresponding angles have degree measures that are represented by the algebraic expressions $9x - 15°$ and $6x + 30°$. Since $l_1 \parallel l_2$, this pair of corresponding angles are congruent.

$9x - 15° = 6x + 30°$	The angle measures are equal.
$3x - 15° = 30°$	To eliminate $6x$ from the right side, subtract $6x$ from both sides.
$3x = 45°$	To isolate the variable term $3x$, undo the subtraction of $15°$ by adding $15°$ to both sides: $30° + 15° = 45°$.
$x = 15°$	To isolate x, undo the multiplication by 3 by dividing both sides by 3.

Thus, x is $15°$.

Self Check 4
In the figure below, $l_1 \parallel l_2$. Find y.

$7y - 14°$

$4y + 10°$

Answer: $8°$

Now Try
Problem 27

EXAMPLE 5   In the figure, $l_1 \parallel l_2$. **a.** Find x.
b. Find the measures of both angles labeled in the figure.

$3x + 20°$

$3x - 80°$

Strategy
We will use the interior angles property to write an equation that mathematically models the situation.

Why
We can then solve the equation to find x.

Solution
a. Because the angles are interior angles on the same side of the transversal, they are supplementary.

$3x - 80° + 3x + 20° = 180°$	The sum of the measures of two supplementary angles is $180°$.
$6x - 60° = 180°$	Combine like terms.
$6x = 240°$	To undo the subtraction of $60°$, add $60°$ to both sides: $180° + 60° = 240°$.
$x = 40°$	To undo the multiplication by 6, divide both sides by 6.

Thus, x is $40°$.

This problem may be solved using a different approach. In the figure on the next page, we see that $\angle 1$ and the angle with measure $3x - 80°$ are corresponding angles.

Self Check 5
In the figure, $l_1 \parallel l_2$.
a. Find x.
b. Find the measures of both angles labeled in the figure.

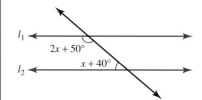

$2x + 50°$

$x + 40°$

Because l_1 and l_2 are parallel, all pairs of corresponding angles are congruent. Therefore,

$$m(\angle 1) = 3x - 80°$$

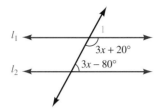

In the figure, we also see that $\angle 1$ and the angle with measure $3x + 20°$ are supplementary. That means that the sum of their measures must be $180°$. We have

$$m(\angle 1) + 3x + 20° = 180°$$

$$3x - 80° + 3x + 20° = 180°   \quad \text{Replace } m(\angle 1) \text{ with } 3x - 80°.$$

This is the same equation that we obtained in the previous solution. When it is solved, we find that x is $40°$.

b. To find the measures of the angles in the figure, we evaluate the expressions $3x + 20°$ and $3x - 80°$ for $x = 40°$.

$$3x + 20° = 3(\mathbf{40}°) + 20° \qquad\qquad 3x - 80° = 3(\mathbf{40}°) - 80°$$
$$= 120° + 20° \qquad\qquad\qquad = 120° - 80°$$
$$= 140° \qquad\qquad\qquad\qquad = 40°$$

The measures of the angles labeled in the figure are $140°$ and $40°$.

Answers: **a.** $30°$ **b.** $110°, 70°$

Now Try
Problem 29

■

4. Use the converse to show that lines are parallel.

Many geometric facts are stated in *if ....., then ....* form. For example, we have seen that

> If two angles are vertical angles, then they are congruent.

When studying such statements, it is worthwhile to interchange their parts and then to determine whether the "reverse" is true.

> If two angles are congruent, then they are vertical angles.

In this case, the resulting statement is not true.

If a mathematical statement is written in the form *if p..., then q...*, we call the statement *if q..., then p...* its **converse.** It is interesting to note that the converses of some statements are true, while the converses of other statements are false.

Earlier in this section, we saw that

- If two parallel lines are cut by a transversal, then corresponding angles are congruent.
- If two parallel lines are cut by a transversal, then alternate interior angles are congruent.

It can be proved that the converses of these two statements are true. That is, given two lines cut by a transversal,

- If a pair of corresponding angles are congruent, then the lines are parallel.
- If a pair of alternate interior angles are congruent, then the lines are parallel.

When a statement and its converse are both true, we can combine them into a single statement using the phrase *if and only if.*

Properties of parallel lines

> Given two lines cut by a transversal,
>
> **1.** Corresponding angles are congruent if and only if the lines are parallel.
>
> **2.** Alternate interior angles are congruent if and only if the lines are parallel.

EXAMPLE 6   In the figure, are lines l_1 and l_2 parallel?

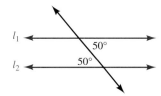

Strategy

We need to identify a pair of corresponding angles or a pair of alternate interior angles that are congruent.

Why

If that is the case, then by a converse statement we can conclude that the lines l_1 and l_2 are parallel.

Solution

In the figure, we have two lines that are cut by a transversal. Because a pair of alternate interior angles are congruent (both have measure 50°), the lines are parallel.

Self Check 6

In the figure below, are lines l_1 and l_2 parallel?

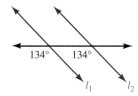

Answer: Since a pair of corresponding angles are congruent, the lines are parallel.

Now Try

Problem 31

STUDY SET   Section 3

VOCABULARY   *Fill in the blanks.*

1. Two lines that lie in the same plane are called _____. Two lines that lie in different planes are called _____.

2. Two coplanar lines that do not intersect are called _____ lines. Two noncoplanar lines that do not intersect are called _____ lines.

3. _____ lines are lines that intersect and form right angles.

4. A line that intersects two coplanar lines in two distinct (different) points is called a(n) _____.

5. In the figure below, ∠4 and ∠6 are _____ interior angles. ∠2 and ∠6 are _____ angles.

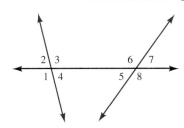

6. If a mathematical statement is written in the form *if p...*, *then q...*, we call the statement *if q..., then p...* its

_____.

CONCEPTS

7. **a.** Draw two parallel lines. Label them l_1 and l_2.

   b. Draw two lines that are not parallel. Label them l_1 and l_2.

8. **a.** Draw two perpendicular lines. Label them l_1 and l_2.

   b. Draw two lines that are not perpendicular. Label them l_1 and l_2.

9. **a.** Draw two parallel lines cut by a transversal. Label the lines l_1 and l_2 and label the transversal l_3.

   b. Draw two lines that are not parallel cut by a transversal. Label the lines l_1 and l_2 and label the transversal l_3.

10. Draw three parallel lines. Label them l_1, l_2, and l_3.

In Problems 11–14, two parallel lines are cut by a transversal. Fill in the blanks.

11. In the figure, ∠*ABC* ≅ ∠*BEF* because when two parallel lines are cut by a transversal, _____ angles are congruent.

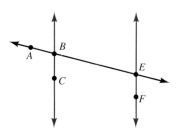

12. In the figure, ∠1 ≅ ∠2 because when two parallel lines are cut by a transversal, _____ angles are congruent.

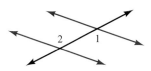

13. In the figure, m(∠ABC) + m(∠BCD) = 180° because when two parallel lines are cut by a transversal, _____ angles on the same side of the transversal are supplementary.

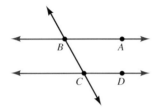

14. In the figure, ∠8 ≅ ∠6 because when two parallel lines are cut by a transversal, _____ angles are congruent.

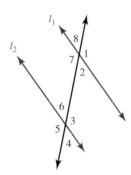

15. In the figure, $l_1 \parallel l_2$. What can you conclude about l_1 and l_3?

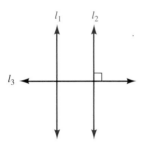

16. In the figure, $l_1 \parallel l_2$ and $l_2 \parallel l_3$. What can you conclude about l_1 and l_3?

NOTATION *Fill in the blanks.*

17. The symbol ⌐ indicates a(n) _____ angle.

18. The symbol ∥ is read as "is _____ to."

19. The symbol ⊥ is read as "is _____ to."

20. The symbol l_1 is read as "line *l* _____ one."

GUIDED PRACTICE

21. Refer to the figure and identify each of the following. *See Example 1.*

    a. corresponding angles

    b. interior angles

    c. alternate interior angles

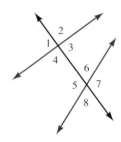

22. Refer to the figure and identify each of the following. *See Example 1.*

    a. corresponding angles

    b. interior angles

    c. alternate interior angles

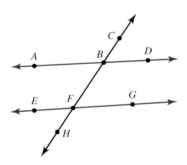

23. In the figure, $l_1 \parallel l_2$ and m(∠4) = 130°. Find the measures of the other seven angles that are labeled. *See Example 2.*

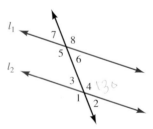

24. In the figure, $l_1 \parallel l_2$ and m(∠2) = 40°. Find the measures of the other angles. *See Example 2.*

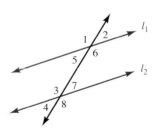

25. In the figure, $\overline{YM} \parallel \overline{XN}$. Which pairs of angles are congruent? ***See Example 3.***

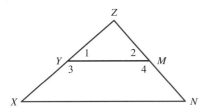

26. In the figure, $\overline{AE} \parallel \overline{BD}$. Which pairs of angles are congruent? ***See Example 3.***

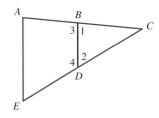

In Problems 27 and 28, $l_1 \parallel l_2$. First find x. Then determine the measure of each angle that is labeled in the figure. ***See Example 4.***

27.

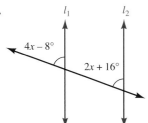

28.

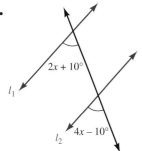

In Problems 29 and 30, $l_1 \parallel l_2$. First find x. Then determine the measure of each angle that is labeled in the figure. ***See Example 5.***

29.

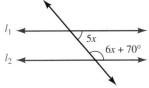

30.

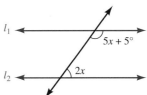

31. Explain why l_1 and l_2 in the figure below are parallel. ***See Example 6.***

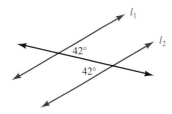

32. Explain why l_1 and l_2 in the figure below are parallel. ***See Example 6.***

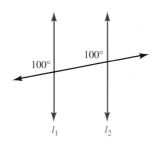

TRY IT YOURSELF

33. In the figure below, $l_1 \parallel AB$. Find
   a. m($\angle 1$), m($\angle 2$), m($\angle 3$), and m($\angle 4$).
   b. m($\angle 3$) + m($\angle 4$) + m($\angle ACD$).
   c. m($\angle 1$) + m($\angle ABC$) + m($\angle 4$).

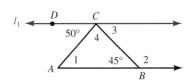

34. In the figure, $\overline{AB} \parallel \overline{DE}$. Find m($\angle B$), m($\angle E$), and m($\angle 1$).

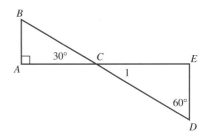

35. In the figure, $\overline{AB} \parallel \overline{DE}$. What pairs of angles are congruent? Explain your reasoning.

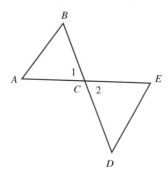

36. In the figure, find the value for x that will make $l_1 \parallel l_2$.

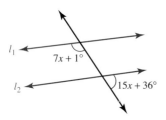

First find x. Then determine the measure of each angle that is labeled in the figure.

37. $l_1 \parallel \overline{CA}$

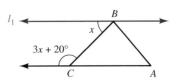

38. $\overline{AB} \parallel \overline{DE}$

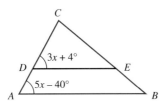

39. $\overline{AB} \parallel \overline{DE}$

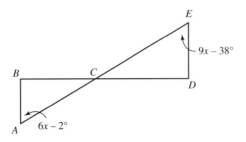

40. $\overline{AC} \parallel \overline{BD}$

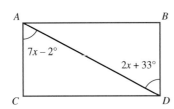

APPLICATIONS

41. CONSTRUCTING PYRAMIDS   The Egyptians used a device called a **plummet** to tell whether stones were properly leveled. A plummet (shown below) is made up of an A-frame and a plumb bob suspended from the peak of the frame. How could a builder use a plummet to tell that the two stones on the left are not level and that the three stones on the right are level?

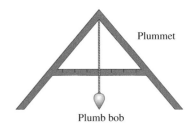

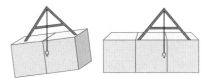

42. DIAGRAMMING SENTENCES   English instructors have their students diagram sentences to help teach proper sentence structure. A diagram of the sentence *The cave was rather dark and damp* is shown below. Point out pairs of parallel and perpendicular lines used in the diagram.

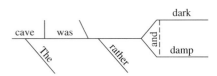

43. LOGO   Point out any perpendicular lines that can be found on the BMW company logo shown below.

44. PAINTING SIGNS   For many sign painters, the most difficult letter to paint is a capital E because of all the right angles involved. How many right angles are there?

45. HANGING WALLPAPER   Explain why the concepts of *perpendicular* and *parallel* are both important when hanging wallpaper.

46. TOOLS   What geometric concepts are seen in the design of the rake shown below?

47. SEISMOLOGY   The figure shows how an earthquake fault occurs when two blocks of earth move apart and one part drops down. Determine the measures of $\angle 1$, $\angle 2$, and $\angle 3$.

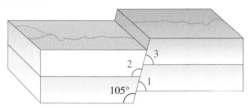

48. CARPENTRY   A carpenter braced three 2 × 4s as shown below and then used a tool to measure the three highlighted angles. If each of those angles measured 45°, what does the carpenter know about the three 2 × 4s? Explain your answer.

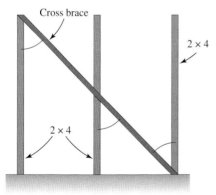

WRITING

49. PARKING DESIGN   Using terms from this section, write a paragraph describing the parking layout shown below.

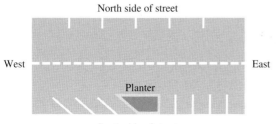

50. In your own words, explain what is meant by each of the following sentences.

   a. The hikers were told that the path *parallels* the river.

   b. John's quick rise to fame and fortune *paralleled* that of his older brother.

   c. The judge stated that the case that was before her court was without *parallel*.

51. Are lines l_1 and l_2 in the figure parallel? Explain your answer.

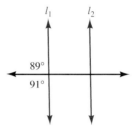

52. Are lines l_1 and l_2 in the figure parallel? Explain why or why not.

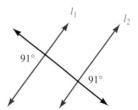

53. Explain why l_1 and l_2 in the figure are not parallel.

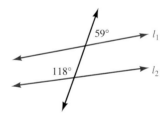

54. Write a true *if … then* statement whose converse is false.

55. Are pairs of alternate interior angles always congruent? Explain.

56. Are pairs of interior angles on the same side of a transversal always supplementary? Explain.

4 *Triangles*

Objectives

1. Classify polygons.
2. Classify triangles.
3. Identify isosceles triangles.
4. Find unknown angle measures of triangles.

We will now discuss geometric figures called *polygons*. We see these shapes every day. For example, the walls of most buildings are rectangular in shape. Some tile and vinyl floor patterns use the shape of a pentagon or a hexagon. Stop signs are in the shape of an octagon.

In this section, we will focus on one specific type of polygon called a *triangle*. Triangular shapes are especially important because triangles contribute strength and stability to walls and towers. The gable roofs of houses are triangular, as are the sides of many ramps.

THE HOUSE OF THE SEVEN GABLES, SALEM, MASSACHUSETTS

1. Classify polygons.

Polygon

> A **polygon** is a closed geometric figure with at least three line segments for its sides.

Polygons are formed by fitting together line segments in such a way that

- no two of the segments intersect, except at their endpoints, and
- no two line segments with a common endpoint lie on the same line.

The line segments that form a polygon are called its **sides.** The point where two sides of a polygon intersect is called a **vertex** of the polygon (plural **vertices**). The polygon shown below has 5 sides and 5 vertices.

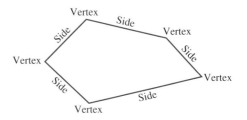

Polygons are classified according to the number of sides that they have. For example, in the figure below, we see that a polygon with four sides is called a *quadrilateral,* and a polygon with eight sides is called an *octagon.* If a polygon has sides that are all the same length and angles that are the same measure, we call it a **regular polygon.**

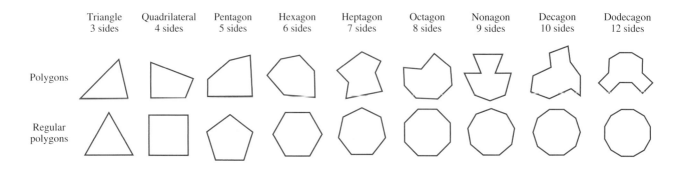

EXAMPLE 1   Give the number of vertices of **a.** a triangle **b.** a hexagon

Strategy

We will determine the number of angles that each polygon has.

Why

The number of its vertices is equal to the number of its angles.

Solution

a. From the figure above, we see that a triangle has three angles and therefore three vertices.

b. From the figure above, we see that a hexagon has six angles and therefore six vertices.

From the results of Example 1, we see that the number of vertices of a polygon is equal to the number of its sides.

2. Classify triangles.

A **triangle** is a polygon with three sides (and three vertices). Recall that in geometry points are labeled with capital letters. We can use the capital letters that denote the vertices of a triangle to name the triangle. For example, when referring to the triangle in the left margin, with vertices *A*, *B*, and *C*, we can use the notation △*ABC* (read as "triangle *ABC*").

CAUTION   When naming a triangle, we may begin with any vertex. Then we move around the figure in a clockwise (or counterclockwise) direction as we list the remaining vertices. Other ways of naming the triangle in the margin are △*ACB*, △*BCA*, △*BAC*, △*CAB*, and △*CBA*.

The figures below show how triangles can be classified according to the lengths of their sides. The single tick marks drawn on each side of the equilateral triangle indicate that the sides are of equal length. The double tick marks drawn on two of the sides of the isosceles triangle indicate that they have the same length. Each side of the scalene triangle has a different number of tick marks to indicate that the sides have different lengths.

Equilateral triangle (all sides equal length)   Isosceles triangle (at least two sides of equal length)   Scalene triangle (no sides of equal length)

CAUTION   Since equilateral triangles have at least two sides of equal length, they are also isosceles. However, isosceles triangles are not necessarily equilateral.

Triangles may also be classified by their angles, as shown below.

Acute triangle (has three acute angles)   Obtuse triangle (has an obtuse angle)   Right triangle (has a right angle)

Right triangles have many real-life applications. For example, in figure (a) on the next page, we see that a right triangle is formed when a ladder leans against the wall of a building.

The longest side of a right triangle is called the **hypotenuse,** and the other two sides are called **legs.** The hypotenuse of a right triangle is always opposite the 90° (right) angle. The legs of a right triangle are adjacent to (next to) the right angle, as shown in figure (b).

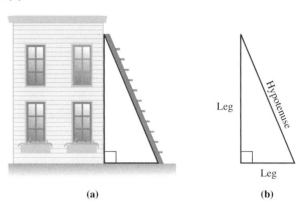

(a)   (b)

3. Identify isosceles triangles.

In an isosceles triangle, the angles opposite the sides of equal length are called **base angles,** the sides of equal length form the **vertex angle,** and the third side is called the **base.** Two examples of isosceles triangles are shown below.

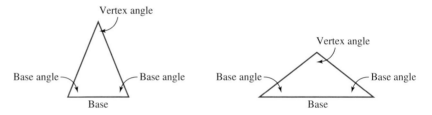

We have seen that isosceles triangles have two sides of equal length. The **isosceles triangle theorem** states that such triangles have one other important characteristic: Their base angles are congruent.

Isosceles triangle theorem

> If two sides of a triangle are congruent, then the angles opposite those sides are congruent.

 CAUTION   Tick marks can be used to denote the sides of a triangle that have the same length. They can also be used to indicate the angles of a triangle with the same measure. For example, we can show that the base angles of the isosceles triangle below are congruent by using single tick marks.

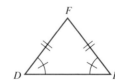

$\angle D$ is opposite $\overline{FE}$, and $\angle E$ is opposite $\overline{FD}$. By the isosceles triangle theorem, if $m(\overline{FD}) = m(\overline{FE})$, then $m(\angle D) = m(\angle E)$.

The *converse* of the isosceles triangle theorem is also true.

Converse of the isosceles triangle theorem

> If two angles of a triangle are congruent, then the sides opposite the angles have the same length, and the triangle is isosceles.

EXAMPLE 2 Is the triangle in the figure an isosceles triangle?

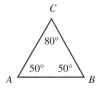

Strategy

We will consider the measures of the angles of the triangle.

Why

If two angles of a triangle are congruent, then the sides opposite the angles have the same length, and the triangle is isosceles.

Solution

∠*A* and ∠*B* have the same measure, 50°. By the converse of the isosceles triangle theorem, if m(∠*A*) = m(∠*B*), we know that m($\overline{BC}$) = m($\overline{AC}$) and that △*ABC* is isosceles.

Self Check 2

Is the triangle shown below an isosceles triangle?

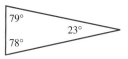

Answer: no

Now Try

Problems 39 and 41

4. Find unknown angle measures of triangles.

To prove a very important fact about triangles, we begin with △*ABC* as shown below. Line *l* is drawn through point *B* so that it is parallel to $\overline{AC}$. For easy reference, several of the angles are labeled with numbers. Three observations from the figure will be helpful in the proof.

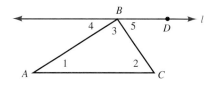

Observation 1: ∠*ABD* is made up of ∠5 and ∠3. Therefore,

m(∠*ABD*) = m(∠5) + m(∠3)

Observation 2: Since the measure of a straight angle is 180°,

m(∠*ABD*) + m(∠4) = 180°

Observation 3: Since $\overleftrightarrow{AB}$ is a transversal cutting parallel lines $\overleftrightarrow{AC}$ and *l*, alternate interior angles are congruent, and therefore

m(∠1) = m(∠4)

Similarly, since $\overleftrightarrow{BC}$ is a transversal cutting parallel lines $\overleftrightarrow{AC}$ and *l*, alternate interior angles are congruent, and therefore

m(∠2) = m(∠5)

Proof To begin the proof, we apply the addition property of equality to the equation from Observation 1.

$$m(\angle ABD) = m(\angle 5) + m(\angle 3)$$
$$\mathbf{m(\angle ABD) + m(\angle 4)} = \mathbf{m(\angle 4)} + m(\angle 5) + m(\angle 3) \quad \text{Add } m(\angle 4) \text{ to both sides.}$$

From Observation 2, we can replace m(∠*ABD*) + m(4) with 180°.

$$180° = \mathbf{m(\angle 4) + m(\angle 5)} + m(\angle 3)$$

From Observation 3, we can replace m(∠4) with m(∠1), and m(∠5) with m(∠2).

$$180° = \mathbf{m(\angle 1) + m(\angle 2)} + m(\angle 3)$$

The result indicates that the sum of the angles of △ABC in the figure is 180°. Because of the generality of this proof, we have shown this fact to be true for any triangle.

Angles of a triangle

| The sum of the angle measures of any triangle is 180°. |

If you draw several triangles and carefully measure each angle with a protractor, you will find that the sum of the angle measures of each triangle is 180°. Two examples are shown below.

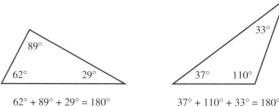

$62° + 89° + 29° = 180°$ $37° + 110° + 33° = 180°$

EXAMPLE 3 In the figure, find *x*.

Strategy
We will use the fact that the sum of the angle measures of any triangle is 180° to write an equation that mathematically models the situation.

Why
We can then solve the equation to find the unknown angle measure, *x*.

Solution
Since the sum of the angle measures of any triangle is 180°, we have

$x + 40° + 90° = 180°$ The ⌐ symbol indicates that the measure of the angle is 90°.

$x + 130° = 180°$ Do the addition: $40° + 90° = 130°$.

$x = 50°$ To isolate *x*, undo the addition of 130° by subtracting 130° from both sides.

Thus, *x* is 50°.

Self Check 3
In the figure, find *y*.

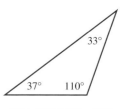

Answer: 90°

Now Try
Problems 43 and 49

For more complicated problems, it is often helpful to use the following five-step problem-solving strategy.

EXAMPLE 4   In △ABC, the measure of ∠A exceeds the measure of ∠B by 32°, and the measure ∠C is twice the measure of ∠B. Find the measure of each angle of △ABC.

Analyze the problem
• There are three angles to consider: ∠A, ∠B, and ∠C.
• The measure of ∠A exceeds the measure of ∠B by 32°.
• The measure of ∠C is twice the measure of ∠B.
• We are to find the measure of each angle.

Form an equation
Since the measures of ∠A and ∠C are related to the measure of ∠B, we begin by letting *x* = the measure of ∠B. Then we translate the words *exceeds by 32°* and *twice*, to write algebraic expressions that represent the measures of the other two angles.

$x + 32°$ = the measure of ∠A

$2x$ = the measure of ∠C

At this stage, drawing a sketch of the situation is often helpful.

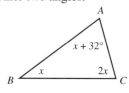

Because the sum of the measures of the angles of any triangle is 180°,

The measure of ∠A	plus	the measure of ∠B	plus	the measure of ∠C	is	180°.
$x + 32°$	+	x	+	$2x$	=	180°

Solve the equation

$$x + 32° + x + 2x = 180°$$
$$4x + 32° = 180° \qquad \text{Combine like terms: } x + x + 2x = 4x.$$
$$4x + 32° - \mathbf{32°} = 180° - \mathbf{32°} \qquad \text{To isolate the variable term, } 4x, \text{ undo the addition of } 32° \text{ by subtracting } 32° \text{ from both sides.}$$
$$4x = 148° \qquad \text{Do the subtractions.}$$
$$\frac{4x}{4} = \frac{148°}{4} \qquad \text{To isolate } x, \text{ undo the multiplication by 4 by dividing both sides by 4.}$$
$$x = 37° \qquad \text{Do the divisions. This is the measure of } \angle B.$$

To find the measures of ∠A and ∠C, we evaluate the expressions $x + 32°$ and $2x$ for $x = 37°$.

$$x + 32° = \mathbf{37°} + 32° \quad \text{Substitute 37 for } x. \qquad 2x = 2(\mathbf{37°}) \quad \text{Substitute 37 for } x.$$
$$= 69° \qquad\qquad\qquad\qquad = 74°$$

State the conclusion   The measure of ∠B is 37°, the measure of ∠A is 69°, and the measure of ∠C is 74°.

Check the result   First, we note that 69° exceeds 37° by 32° and that 74° is twice 37°. Furthermore, since $37° + 69° + 74° = 180°$, the results check.

Self Check 4

In △*DEF*, the measure of ∠D exceeds the measure of ∠E by 5°, and the measure of ∠F is three times the measure of ∠E. Find the measure of each angle of △*DEF*.

Answer: m∠E = 35°, m∠D = 40°, m∠F = 105°

Now Try

Problem 51

EXAMPLE 5   If one base angle of an isosceles triangle measures 70°, what is the measure of the vertex angle?

Strategy

We will use the isosceles triangle theorem and the fact that the sum of the angle measures of any triangle is 180° to write an equation that mathematically models the situation.

Why

We can then solve the equation to find the unknown angle measure.

Solution

By the isosceles triangle theorem, if one of the base angles measures 70°, so does the other. (See the figure on the right.) If we let x repre-sent the measure of the vertex angle, we have

$$x + 70° + 70° = 180° \quad \text{The sum of the measures of the angles of a triangle is 180°.}$$
$$x + 140° = 180° \quad \text{Combine like terms: } 70° + 70° = 140°.$$
$$x = 40° \quad \text{To isolate } x, \text{ undo the addition of } 140° \text{ by subtracting } 140° \text{ from both sides.}$$

The vertex angle measures 40°.

Self Check 5

If one base angle of an isosceles triangle measures 33°, what is the measure of the vertex angle?

Answer: 114°

Now Try

Problem 55

EXAMPLE 6   If the vertex angle of an isosceles triangle measures 99°, what are the measures of the base angles?

Strategy
We will use the fact that the base angles of an isosceles triangle have the same measure and the sum of the angle measures of any triangle is 180° to write an equation that mathematically models the situation.

Why
We can then solve the equation to find the unknown angle measures.

Solution

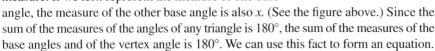

The base angles of an isosceles triangle have the same measure. If we let x represent the measure of one base angle, the measure of the other base angle is also x. (See the figure above.) Since the sum of the measures of the angles of any triangle is 180°, the sum of the measures of the base angles and of the vertex angle is 180°. We can use this fact to form an equation.

$$x + x + 99° = 180°$$

$2x + 99° = 180°$ Combine like terms: $x + x = 2x$.

$2x = 81°$ To isolate the variable term, $2x$, undo the addition of 99° by subtracting 99° from both sides.

$\dfrac{2x}{2} = \dfrac{81°}{2}$ To isolate x, undo the multiplication by 2 by dividing both sides by 2.

$x = 40.5°$

The measure of each base angle is 40.5°.

Self Check 6
If the vertex angle of an isosceles triangle measures 57°, what are the measures of the base angles?

Answer: 61.5°

Now Try
Problem 59

STUDY SET   Section 4

VOCABULARY   *Fill in the blanks.*

1. A _____ is a closed geometric figure with at least three line segments for its sides.

2. The polygon shown below has seven _____ and seven vertices.

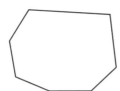

3. A point where two sides of a polygon intersect is called a _____ of the polygon.

4. A _____ polygon has sides that are all the same length and angles that all have the same measure.

5. A triangle with three sides of equal length is called an _____ triangle. An _____ triangle has at least two sides of equal length. A _____ triangle has no sides of equal length.

6. An _____ triangle has three acute angles. An _____ triangle has one obtuse angle. A _____ triangle has one right angle.

7. The longest side of a right triangle is called the _____. The other two sides of a right triangle are called _____.

8. The _____ angles of an isosceles triangle have the same measure. The sides of equal length of an isosceles triangle form the _____ angle.

9. In this section, we discussed the sum of the measures of the angles of a triangle. The word *sum* indicates the operation of _____.

10. Complete the table.

Number of sides	Name of polygon
3	
4	
5	
6	
7	
8	
9	
10	
12	

CONCEPTS

11. Draw an example of each type of regular polygon.

    a. hexagon   **b.** octagon

    c. quadrilateral   **d.** triangle

    e. pentagon   **f.** decagon

12. Explain why the given polygon is not a regular polygon.

    a.   **b.**

13. Refer to the triangle below.

    a. What are the names of the vertices of the triangle?

    b. How many sides does the triangle have? Name them.

    c. Use the vertices to name this triangle in three ways.

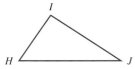

14. Draw an example of each type of triangle.

    a. isosceles   **b.** equilateral

    c. scalene

15. Draw an example of each type of triangle.

    a. obtuse   **b.** right

    c. acute

16. Classify each triangle as an acute, an obtuse, or a right triangle.

    a. 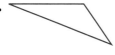   **b.**

    c. 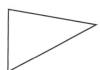   **d.**

17. Refer to the triangle shown below.

    a. What is the measure of ∠B?

    b. What type of triangle is it?

    c. What two line segments form the legs?

    d. What line segment is the hypotenuse?

    e. Which side of the triangle is the longest?

    f. Which side is opposite ∠B?

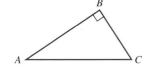

18. Fill in the blanks.

    a. The sides of a right triangle that are adjacent to the right angle are called the _____.

    b. The hypotenuse of a right triangle is the side _____ the right angle.

19. Fill in the blanks.

    a. The _____ triangle theorem states that if two sides of a triangle are congruent, then the angles opposite those sides are congruent.

    b. The _____ of the isosceles triangle theorem states that if two angles of a triangle are congruent, then the sides opposite the angles have the same measure, and the triangle is isosceles.

20. Refer to the triangle below.

    a. What two sides are of equal length?

    b. What type of triangle is △XYZ?

    c. Name the base angles.

    d. Which side is opposite ∠X?

    e. What is the vertex angle?

    f. Which angle is opposite side $\overline{XY}$?

    g. Which two angles are congruent?

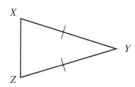

21. Refer to the triangle below.

    a. What do we know about $\overline{EF}$ and $\overline{GF}$?

    b. What type of triangle is △EFG?

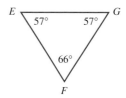

22. a. Find the sum of the measures of the angles of △JKL, shown in figure (a).

    b. Find the sum of the measures of the angles of △CDE, shown in figure (b).

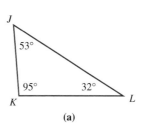   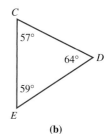

         (a)   **(b)**

23. What is the sum of the measures of the angles of any triangle?

24. Suppose we let x = the measure of $\angle B$.

    a. If the measure of $\angle C$ is twice the measure of $\angle B$, write an algebraic expression that represents the measure of $\angle C$.

    b. If the measure of $\angle D$ is 5° less than four times the measure of $\angle B$, write an algebraic expression that represents the measure of $\angle D$.

NOTATION   *Fill in the blanks.*

25. The symbol $\triangle$ means _____.

26. The symbol m($\angle A$) means the _____ of angle A.

27. The symbol m($\overline{AB}$) means the measure of line _____ AB.

28. The symbol $\llcorner$ indicates a _____ angle is formed.

Refer to the triangle below.

29. What fact about the sides of $\triangle ABC$ do the tick marks indicate?

30. What fact about the angles of $\triangle ABC$ do the tick marks indicate?

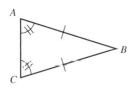

GUIDED PRACTICE   *For each polygon, give the number of sides it has, give its name, and then give the number of vertices that it has.* **See Example 1.**

31. a.   **b.**

32. a.   **b.**

33. a.   **b.**

34. a.   **b.**

Classify each triangle as an equilateral triangle, an isosceles triangle, or a scalene triangle. **See Objective 2.**

35. a. 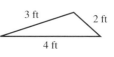   **b.**

36. a.   **b.**

37. a.   **b.**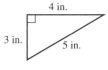

38. a. 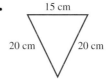   **b.**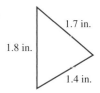

State whether each of the triangles is an isosceles triangle. **See Example 2.**

39. 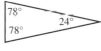   **40.**

41.

42.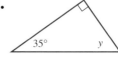

Find y. **See Example 3.**

43.    **44.**

45.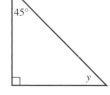

46.

The degree measures of the angles of a triangle are represented by algebraic expressions. First find x. Then determine the measure of each angle of the triangle. **See Example 4.**

47.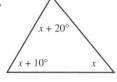

48.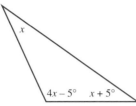

49.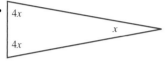

50.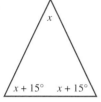

Find the measure of each angle of △ABC given the following information. **See Example 4.**

51. The measure of ∠A exceeds the measure of ∠B by 12°, and the measure of ∠C is twice the measure of ∠B.

52. The measure of ∠A exceeds the measure of ∠B by 24°, and the measure of ∠C is twice the measure of ∠B.

53. The measure of ∠B is 45° more than the measure of ∠C, and the measure of ∠A is 45° less than the measure of ∠C.

54. The measure of ∠B is 11° less than the measure of ∠C, and the measure of ∠A is 61° less than the measure of ∠C.

Find the measure of the vertex angle of each isosceles triangle given the following information. **See Example 5.**

55. The measure of one base angle is 56°.

56. The measure of one base angle is 68°.

57. The measure of one base angle is 85.5°.

58. The measure of one base angle is 4.75°.

Find the measure of one base angle of each isosceles triangle given the following information. **See Example 6.**

59. The measure of the vertex angle is 102°.

60. The measure of the vertex angle is 164°.

61. The measure of the vertex angle is 90.5°.

62. The measure of the vertex angle is 2.5°.

TRY IT YOURSELF

Find the measure of each vertex angle.

63.

64.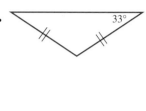

65.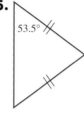

66.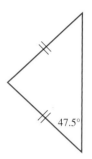

The measures of two angles of △ABC are given. Find the measure of the third angle.

67. m($\angle A$) = 30° and m($\angle B$) = 60°; find m($\angle C$).

68. m($\angle A$) = 45° and m($\angle C$) = 105°; find m($\angle B$).

69. m($\angle B$) = 100° and m($\angle A$) = 35°; find m($\angle C$).

70. m($\angle B$) = 33° and m($\angle C$) = 77°; find m($\angle A$).

71. m($\angle A$) = 25.5° and m($\angle B$) = 63.8°; find m($\angle C$).

72. m($\angle B$) = 67.25° and m($\angle C$) = 72.5°; find m($\angle A$).

73. m($\angle A$) = 29° and m($\angle C$) = 89.5°; find m($\angle B$).

74. m($\angle A$) = 4.5° and m($\angle B$) = 128°; find m($\angle C$).

Find x.

75.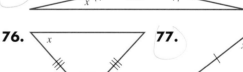

76.

77.

78.

79. One angle of an isosceles triangle has a measure of 39°. What are the possible measures of the other angles?

80. One angle of an isosceles triangle has a measure of 2°. What are the possible measures of the other angles?

81. The measure of each base angle of an isosceles triangle is 10° less than twice the measure of the vertex angle. Find the measure of each angle of the triangle.

82. The measure of a base angle of an isosceles triangle is 5° more than eight times the measure of the vertex angle. Find the measure of each angle of the triangle.

83. The measure of one angle of a triangle is twice that of another angle. The third angle measures 60°. Find the measure of each angle of the triangle.

84. The measure of one angle of a triangle is three times that of another angle. The third angle measures 80°. Find the measure of each angle of the triangle.

85. In $\triangle ABC$, the measure of $\angle A$ is 10° more than twice the measure of $\angle B$. The measure of $\angle C$ is 10° less than the measure of $\angle B$. Find the measure of each angle of the triangle.

86. In $\triangle ABC$, the measure of $\angle A$ is 2° less than three times the measure of $\angle B$. The measure of $\angle C$ is 7° more than the measure of $\angle B$. Find the measure of each angle of the triangle.

87. Find m($\angle C$).

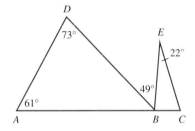

88. Find
   a. m($\angle MXZ$)
   b. m($\angle MYN$)

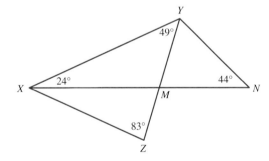

89. Find m($\angle NOQ$).

90. Find m($\angle S$).

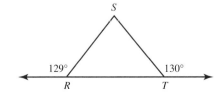

APPLICATIONS

91. POLYGONS IN NATURE   As seen below, a starfish fits the shape of a pentagon. What polygon shape do you see in each of the other objects? **a.** lemon   **b.** chili pepper   **c.** apple

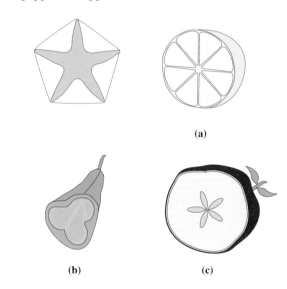

(a)

(b)   (c)

92. CHEMISTRY   Polygons are used to represent the chemical structure of compounds graphically. In the figure below, what types of polygons are used to represent methylprednisolone, the active ingredient in an anti-inflammatory medication?

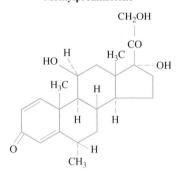

93. AUTOMOBILE JACK   Refer to the figure below. Show that no matter how high the jack is raised, it always forms two isosceles triangles.

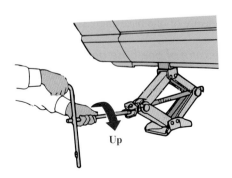

94. EASELS   Refer to the figure below. What type of triangle studied in this section is used in the design of the legs of the easel?

95. POOL   The rack shown below is used to set up the billiard balls when beginning a game of pool. Although it does not meet the strict definition of a polygon, the rack has a shape much like a type of triangle discussed in this section. Which type of triangle?

96. DRAFTING   Among the tools used in drafting are the two clear plastic triangles shown below. Classify each according to the lengths of its sides and then according to its angle measures.

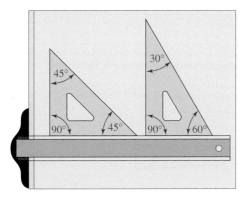

WRITING

97. In this section, we discussed the definition of a pentagon. What is *the* Pentagon? Why is it named that?

98. A student cut a triangular shape out of blue construction paper and labeled the angles ∠1, ∠2, and ∠3, as shown in figure (a) below. Then she tore off each of the three corners and arranged them as shown in figure (b). Explain what important geometric concept this model illustrates.

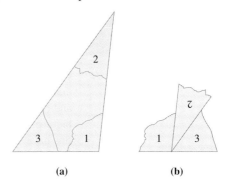

(a)   (b)

99. Explain why a triangle cannot have two right angles.

100. Explain why a triangle cannot have two obtuse angles.

5 *Quadrilaterals and Other Polygons*

Objectives

1. Classify quadrilaterals.
2. Use properties of rectangles to find unknown angle measures and side lengths.
3. Find unknown angle measures of trapezoids.
4. Use the formula for the sum of the angle measures of a polygon.
5. Find unknown angle measures of regular polygons.
6. Find the number of sides of a regular polygon given the measure of an angle.
7. Find the measures of exterior angles of regular polygons.

Recall from Section 4 that a polygon is a closed geometric figure with at least three line segments for its sides. In this section, we will focus on polygons with four sides, called *quadrilaterals*. One type of quadrilateral is the *square*. The game boards for Monopoly and Scrabble have a square shape. Another type of quadrilateral is the *rectangle*. Most picture frames and many mirrors are rectangular. Utility knife blades and swimming fins have shapes that are examples of a third type of quadrilateral called a *trapezoid*.

1. Classify quadrilaterals.

A **quadrilateral** is a polygon with four sides. Some common quadrilaterals are shown below.

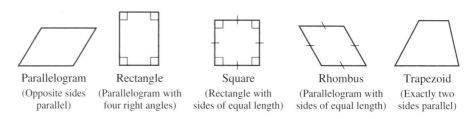

Parallelogram	Rectangle	Square	Rhombus	Trapezoid
(Opposite sides parallel)	(Parallelogram with four right angles)	(Rectangle with sides of equal length)	(Parallelogram with sides of equal length)	(Exactly two sides parallel)

We can use the capital letters that label the vertices of a quadrilateral to name it. For example, when referring to the quadrilateral shown below, with vertices *A*, *B*, *C*, and *D*, we can use the notation quadrilateral *ABCD*.

CAUTION   When naming a quadrilateral (or any other polygon), we may begin with any vertex. Then we move around the figure in a clockwise (or counterclockwise) direction as we list the remaining vertices. Some other ways of naming the quadrilateral above are quadrilateral *ADCB*, quadrilateral *CDAB*, and quadrilateral *DABC*. It would be unacceptable to name it as quadrilateral *ACDB*, because the vertices would not be listed in clockwise (or counterclockwise) order.

A segment that joins two nonconsecutive vertices of a polygon is called a **diagonal** of the polygon. The quadrilateral shown below has two diagonals, $\overline{AC}$ and $\overline{BD}$.

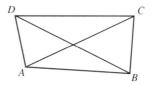

2. Use properties of rectangles to find unknown angle measures and side lengths.

Recall that a **rectangle** is a quadrilateral with four right angles. The rectangle is probably the most common and recognizable of all geometric figures. For example, most doors and windows are rectangular in shape. The boundaries of football fields, soccer fields, and basketball courts are rectangles. Even our paper currency, such as the $1, $5, and $20 bills, is in the shape of a rectangle. Rectangles have several important characteristics.

Properties of rectangles

> In any rectangle:
> 1. All four angles are right angles.
> 2. Opposite sides are parallel.
> 3. Opposite sides have equal length.
> 4. The diagonals have equal length.
> 5. The diagonals intersect at their midpoints.

EXAMPLE 1 In the figure, quadrilateral *WXYZ* is a rectangle. Find each measure: **a.** m(∠*YXW*), **b.** m($\overline{XY}$), **c.** m($\overline{WY}$), and **d.** m($\overline{XZ}$).

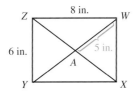

Strategy
We will use properties of rectangles to find the unknown angle measure and the unknown measures of the line segments.

Why
Quadrilateral *WXYZ* is a rectangle.

Solution
a. In any rectangle, all four angles are right angles. Therefore, ∠*YXW* is a right angle, and m(∠*YXW*) = 90°.
b. $\overline{XY}$ and $\overline{WZ}$ are opposite sides of the rectangle, so they have equal length. Since the length of $\overline{WZ}$ is 8 inches, m($\overline{XY}$) is also 8 inches.
c. $\overline{WY}$ and $\overline{ZX}$ are diagonals of the rectangle, and they intersect at their midpoints. That means that point *A* is the midpoint of $\overline{WY}$. Since the length of $\overline{WA}$ is 5 inches, m($\overline{WY}$) is 2 · 5 inches, or 10 inches.
d. The diagonals of a rectangle are of equal length. In part c, we found that the length of $\overline{WY}$ is 10 inches. Therefore, m($\overline{XZ}$) is also 10 inches.

Self Check 1

In rectangle *RSTU* shown below, the length of $\overline{RT}$ is 13 ft. Find each measure: **a.** m(∠*SRU*), **b.** m($\overline{ST}$), **c.** m($\overline{TG}$), and **d.** m($\overline{SG}$).

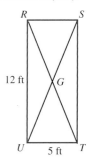

Answers: **a.** 90° **b.** 12 ft **c.** 6.5 ft **d.** 6.5 ft

Now Try
Problem 29

We have seen that if a quadrilateral has four right angles, it is a rectangle. The following theorems establish some conditions that a parallelogram must meet to ensure that it is a rectangle.

Parallelogram theorems

> 1. If a parallelogram has one right angle, then the parallelogram is a rectangle.
> 2. If the diagonals of a parallelogram are congruent, then the parallelogram is a rectangle.

EXAMPLE 2 **Construction.** A carpenter wants to build a shed with a 9-foot-by-12-foot base. How can he make sure that the foundation has four right-angle corners?

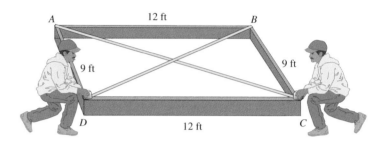

Strategy

The carpenter should find the lengths of the diagonals of the foundation.

Why

If the diagonals are congruent, then the foundation is rectangular in shape and the corners are right angles.

Solution

The four-sided foundation, which we will label as quadrilateral *ABCD*, has opposite sides of equal length. The carpenter can use a tape measure to find the lengths of the diagonals $\overline{AC}$ and $\overline{BD}$. If these diagonals are of equal length, the foundation will be a rectangle and have right angles at its four corners. This process is commonly referred to as "squaring a foundation." Picture framers use a similar process to make sure their frames have four 90° corners.

Now Try

Problem 93

3. Find unknown angle measures of trapezoids.

A **trapezoid** is a quadrilateral with exactly two sides parallel. For the trapezoid shown below, the parallel sides $\overline{AB}$ and $\overline{DC}$ are called **bases.** To distinguish between the two bases, we will refer to $\overline{AB}$ as the **upper base** and $\overline{DC}$ as the **lower base.** The angles on either side of the upper base are called **upper base angles,** and the angles on either side of the lower base are called **lower base angles.** The nonparallel sides are called **legs.**

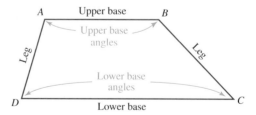

In the figure above, we can view $\overleftrightarrow{AD}$ as a transversal cutting the parallel lines $\overleftrightarrow{AB}$ and $\overleftrightarrow{DC}$. Since $\angle A$ and $\angle D$ are interior angles, they are supplementary. Similarly, $\overleftrightarrow{BC}$ is a transversal cutting the parallel lines $\overleftrightarrow{AB}$ and $\overleftrightarrow{DC}$. Since $\angle B$ and $\angle C$ are interior angles, they are also supplementary. These observations lead us to the conclusion that *there are always two pairs of supplementary angles in any trapezoid.*

EXAMPLE 3   Refer to trapezoid *KLMN* below, with $\overline{KL} \parallel \overline{NM}$. Find *x* and *y*.

Strategy
We will use the interior angles property twice to write two equations that mathematically model the situation.

Why
We can then solve the equations to find *x* and *y*.

Solution
$\angle K$ and $\angle N$ are interior angles on the same side of transversal $\overleftrightarrow{KN}$ that cuts the parallel lines segments $\overline{KL}$ and $\overline{MN}$. Similarly, $\angle L$ and $\angle M$ are interior angles on the same side of transversal $\overleftrightarrow{LM}$ that cuts the parallel lines segments $\overline{KL}$ and $\overline{MN}$. Recall that if two parallel lines are cut by a transversal, interior angles on the same side of the transversal are supplementary. We can use this fact twice—once to find *x* and a second time to find *y*.

$m(\angle K) + m(\angle N) = 180°$ The sum of the measures of supplementary angles is 180°.

$x + 82° = 180°$ Substitute *x* for m($\angle K$) and 82° for m($\angle N$).

$x = 98°$ To isolate *x*, subtract 82° from both sides.

Thus, *x* is 98°.

$m(\angle L) + m(\angle M) = 180°$ The sum of the measures of supplementary angles is 180°.

$121° + y = 180°$ Substitute 121° for m($\angle L$) and *y* for m($\angle M$).

$y = 59°$ To isolate *y*, subtract 121° from both sides.

Thus, *y* is 59°.

Self Check 3
Refer to trapezoid *HIJK* below, with $\overline{HI} \parallel \overline{KJ}$. Find *x* and *y*.

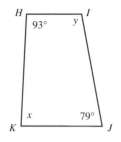

Answer: 87°, 101°

Now Try
Problem 31   ■

If the nonparallel sides of a trapezoid are the same length, it is called an **isosceles trapezoid.** The following figure shows isosceles trapezoid *DEFG* with $\overline{DG} \cong \overline{EF}$. In an isosceles trapezoid, *both pairs of base angles are congruent.* In the figure, $\angle D \cong \angle E$ and $\angle G \cong \angle F$.

EXAMPLE 4   *Landscaping.*   A cross section of a drainage ditch shown below is an isosceles trapezoid with $\overline{AB} \parallel \overline{DC}$. Find *x* and *y*.

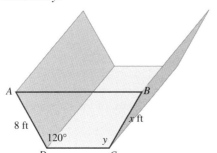

Self Check 4
Refer to the isosceles trapezoid shown below. Find *x* and *y*.

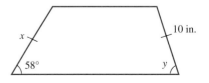

Strategy

We will compare the nonparallel sides and compare a pair of base angles of the trapezoid to find each unknown.

Why

The nonparallel sides of an isosceles trapezoid have the same length and both pairs of base angles are congruent.

Solution

Since $\overline{AD}$ and $\overline{BC}$ are the nonparallel sides of an isosceles triangle, $m(\overline{AD})$ and $m(\overline{BC})$ are equal, and x is 8 ft.

Since $\angle D$ and $\angle C$ are a pair of base angles of an isosceles trapezoid, they are congruent and $m(\angle D) = m(\angle C)$. Thus, y is 120°.

Answer: 10 in., 58°

Now Try

Problem 33 ◼

4. Use the formula for the sum of the angle measures of a polygon.

In the figure shown below, a protractor was used to find the measure of each angle of the quadrilateral. When we add the four angle measures, the result is 360°.

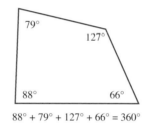

$$88° + 79° + 127° + 66° = 360°$$

This illustrates an important fact about quadrilaterals: The sum of the measures of the angles of *any* quadrilateral is 360°. This can be shown using the diagram in figure (a) below. In the figure, the quadrilateral is divided into two triangles. Since the sum of the angle measures of any triangle is 180°, the sum of the measures of the angles of the quadrilateral is 2 · 180°, or 360°.

A similar approach can be used to find the sum of the measures of the angles of any pentagon or any hexagon. The pentagon in figure (b) is divided into three triangles. The sum of the measures of the angles of the pentagon is 3 · 180°, or 540°. The hexagon in figure (c) is divided into four triangles. The sum of the measures of the angles of the hexagon is 4 · 180°, or 720°. In general, a polygon with *n* sides can be divided into *n* − 2 triangles. Therefore, the sum of the angle measures of a polygon can be found by multiplying 180° by *n* − 2.

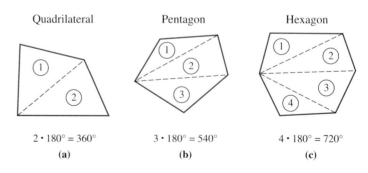

Quadrilateral	Pentagon	Hexagon
2 · 180° = 360°	3 · 180° = 540°	4 · 180° = 720°
(a)	(b)	(c)

Sum of the angles of a polygon

The sum S, in degrees, of the measures of the angles of a polygon with n sides is given by the formula

$$S = (n - 2)180°$$

EXAMPLE 5   Find the sum of the angle measures of a 13-sided polygon.

Strategy
We will substitute 13 for n in the formula $S = (n - 2)180°$ and evaluate the right side.

Why
The variable S represents the unknown sum of the measures of the angles of the polygon.

Solution

$S = (n - 2)180°$ This is the formula for the sum of the measures of the angles of the polygon.

$S = (13 - 2)180°$ Substitute 13 for n, the number of sides.

$\quad = (11)180°$ Do the subtraction within the parentheses.

$\quad = 1,980°$ Do the multiplication.

The sum of the measures of the angles of a 13-sided polygon is $1,980°$.

Self Check 5
Find the sum of the angle measures of the polygon shown below.

Answer: 900°

Now Try
Problem 35

EXAMPLE 6   The sum of the measures of the angles of a polygon is $1,080°$. Find the number of sides the polygon has.

Strategy
We will substitute $1,080°$ for S in the formula $S = (n - 2)180°$ and solve for n.

Why
The variable n represents the unknown number of sides of the polygon.

Solution

$S = (n - 2)180°$ This is the formula for the sum of the measures of the angles of the polygon.

$1,080° = (n - 2)180°$ Substitute $1,080°$ for S, the sum of the measures of the angles.

$1,080° = 180°n - 360°$ Distribute the multiplication by 180°.

$1,080° + 360° = 180°n - 360° + 360°$ To isolate the variable term, $180°n$, undo the subtraction of 360° by adding 360° to both sides.

$1,440° = 180°n$ Do the additions.

$\dfrac{1,440°}{180°} = \dfrac{180°n}{180°}$ To isolate the variable n, undo the multiplication by 180° by dividing both sides by 180°.

$8 = n$ Do the division: $\frac{1,440°}{180°} = 8$.

The polygon has 8 sides. It is an octagon.

Self Check 6
The sum of the measures of the angles of a polygon is $1,620°$. Find the number of sides the polygon has.

Answer: 11

Now Try
Problem 43

---

5. Find unknown angle measures of regular polygons.

Recall that a polygon that has all sides the same length and all angles the same measure is called a *regular polygon*. Two regular polygons are shown on the next page.

Regular hexagon   Regular decagon

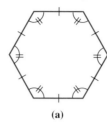   

(a)   (b)

We can find the measure of one angle of a regular polygon using the following formula.

Angle measure of a regular polygon

The angle measure A, in degrees, of one angle of a regular polygon with n sides is given by the formula

$$A = \frac{(n - 2)180°}{n}$$

EXAMPLE 7   Find the measure of one angle of a regular hexagon.

Strategy

A regular hexagon has six angles with the same measure. To find the measure of one of its angles, we will substitute 6 for n in the formula $A = \frac{(n - 2)180°}{n}$ and evaluate the right side.

Why

The variable A represents the unknown measure of one angle of the regular polygon.

Solution

$A = \frac{(n - 2)180°}{n}$ This is the formula for the measure of one angle of a regular polygon.

$A = \frac{(6 - 2)180°}{6}$ Because a regular hexagon has six sides, substitute 6 for n.

$A = \frac{(4)180°}{6}$ Do the subtraction within the parentheses.

$A = \frac{720°}{6}$ Do the multiplication in the numerator.

$A = 120°$ Do the division.

The measure of one angle of a regular hexagon is 120°.

Self Check 7

Find the measure of one angle of a regular decagon.

Answer: 144°

Now Try

Problem 51

6. Find the number of sides of a regular polygon given the measure of an angle.

When we know the measure of one angle of a regular polygon, we can use the following formula to find the number of sides the polygon has.

Regular polygons

The number of sides n that a regular polygon has is given by the formula

$$n = \frac{360°}{180° - A}$$

where A is the measure, in degrees, of one angle of the polygon.

This formula is obtained by solving $A = \frac{(n - 2)180°}{n}$ for n.

EXAMPLE 8   The measure of each angle of a regular polygon is 150°. How many sides does it have?

Strategy

We will find the number of sides of the regular polygon by substituting 150° for A in the formula $n = \frac{360°}{180° - A}$ and evaluating the right side.

Why

The variable n represents the unknown number of sides of the regular polygon.

Solution

$$n = \frac{360°}{180° - A}$$ This is the formula for the number of sides of a regular polygon.

$$n = \frac{360°}{180° - \mathbf{150°}}$$ Substitute 150° for A, the measure of one angle of the regular polygon.

$$= \frac{360°}{30°}$$ Do the subtraction.

$$= 12$$ Do the division.

The polygon has 12 sides.

7. Find the measures of exterior angles of regular polygons.

The pentagon *ABCDE* shown below has five **interior angle**s that are highlighted in red. The five angles that are highlighted in blue are called **exterior angle**s of the polygon.

Exterior angle of a polygon | An **exterior angle** of a polygon is an angle that is adjacent and supplementary to an interior angle.

In the figure, $\angle FAB$ is an exterior angle of the polygon because it is adjacent and supplementary to $\angle BAE$.

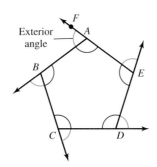

Suppose pentagon *ABCDE* in the figure above is a regular polygon. To find the measure A of one of its five congruent interior angles, we proceed as follows.

$$A = \frac{(n - 2)180°}{n}$$ This is the formula for the measure of an interior angle of a regular polygon.

$$A = \frac{(5 - 2)180°}{5}$$ Because the regular pentagon has five sides, substitute 5 for n.

$$A = \frac{(3)180°}{5}$$ Do the subtraction within the parentheses.

$$= \frac{540°}{5}$$ Do the multiplication.

$$= 108°$$ Do the division.

The measure of each interior angle of the regular pentagon *ABCDE* is 108°. Since each exterior angle of the pentagon is supplementary to an interior angle, each exterior angle has measure 180° − 108° = 72°. For example, m(∠*BAE*) = 108°, and m(∠*FAB*) = 180° − 108° = 72°.

Measure of an exterior angle of a regular polygon	If *A* is the measure of an interior angle and *E* is the measure of an exterior angle of a regular polygon, then $$E = 180° - A$$

EXAMPLE 9   Find the measure of an exterior angle of a regular hexagon.

Strategy
We will find the measure of an exterior angle of a regular hexagon by substituting for *A* in the formula $E = 180° - A$ and evaluating the right side.

Why
The variable *E* represents the unknown measure of the exterior angle.

Solution
Recall from Example 7 that the measure of one interior angle of a regular hexagon is 120°.

$E = 180° - A$ This is the formula for the measure of an exterior angle of a regular polygon.

$E = 180° - \textbf{120°}$ Because the measure of an interior angle of a regular pentagon is 120°, substitute 120° for *A*.

$E = 60°$ Do the subtraction.

The measure of an exterior angle of a regular hexagon is 60°.

Self Check 9
Find the measure of an exterior angle of a regular decagon. (*Hint:* See Self Check 7.)

Answer: 36°

Now Try
Problem 67

■

We can derive another formula for the measure of an exterior angle of a regular polygon by replacing *A*, the measure of an interior angle, with $\frac{(n-2)180°}{n}$ and simplifying.

$E = 180° - A$

$E = 180° - \left[\frac{(n-2)\,180°}{n} \right]$

$= 180° - \left[\frac{180°n - 360°}{n} \right]$ In the numerator, distribute the multiplication by 180°.

$= 180° - \left[\frac{180°n}{n} - \frac{360°}{n} \right]$ Within the brackets: Divide each term of the numerator by *n*.

$= 180° - 180° + \frac{360°}{n}$ In the first fraction, remove the common factor of *n* in the numerator and denominator.

$= \frac{360°}{n}$ Simplify by combing like terms: 180° − 180° = 0.

Measure of an exterior angle of a polygon	If *E* is the measure of an exterior angle of a regular polygon with *n* sides, then $$E = \frac{360°}{n}$$

EXAMPLE 10   Find the measure of an exterior angle of a 15-sided regular polygon.

Strategy
We will find the measure of an exterior angle of a 15-sided regular polygon by substituting for *n* in the formula $E = \frac{360°}{n}$ and evaluating the right side.

Why
The variable *E* represents the unknown measure of the exterior angle.

Solution

$E = \dfrac{360°}{n}$ This is the formula for the measure of an exterior angle of a regular polygon.

$E = \dfrac{360°}{15}$ Substitute 15 for *n*, the number of sides of the regular polygon.

$E = 24°$ Do the division.

The measure of an exterior angle of a 15-sided regular polygon is 24°.

Self Check 10
Find the measure of an exterior angle of a 20-sided regular polygon.

Answer: 18°

Now Try
Problem 75   ■

Just as it is useful to know the sum of the measures of the interior angles of a regular polygon, it is also helpful to know the sum of the measures of its exterior angles. Since there are *n* angles in a regular polygon with *n* sides, the sum *S* of the measures of the exterior angles, one at each vertex, of a regular polygon is the product of the number of sides *n* of the polygon and the measure *E* of one exterior angle.

$S = n \cdot E$

$S = n \cdot \dfrac{360°}{n}$ Since $E = \frac{360°}{n}$, replace *E* with $\frac{360°}{n}$.

$S = 360°$ Remove the common factor of *n* in the numerator and denominator.

This result proves the following property of exterior angles.

Sum of the exterior angles of a regular polygon

> The sum of the measures of the exterior angles of a regular polygon is 360°.

STUDY SET   Section 5

VOCABULARY   *Fill in the blanks.*

1. A _____ is a polygon with four sides.
2. A _____ is a quadrilateral with opposite sides parallel.
3. A _____ is a quadrilateral with four right angles.
4. A rectangle with all sides of equal length is a _____.
5. A _____ is a parallelogram with four sides of equal length.
6. A segment that joins two nonconsecutive vertices of a polygon is called a _____ of the polygon.
7. A _____ has two sides that are parallel and two sides that are not parallel. The parallel sides are called

_____. The legs of an _____ trapezoid have the same length.

8. A _____ polygon has sides that are all the same length and angles that are all the same measure.

CONCEPTS

9. Refer to the polygon on the next page.
   a. How many vertices does it have? List them.
   b. How many sides does it have? List them.
   c. How many diagonals does it have? List them.

d. Tell which of the following are acceptable ways of naming the polygon.

quadrilateral *ABCD*
quadrilateral *CDBA*
quadrilateral *ACBD*
quadrilateral *BADC*

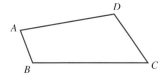

10. Draw an example of each type of quadrilateral.

a. rhombus   **b.** parallelogram

c. trapezoid   **d.** square

e. rectangle   **f.** isosceles trapezoid

11. A parallelogram is shown below. Fill in the blanks.

a. $\overline{ST}\,\|$ _____   **b.** $\overline{SV}$ ___ $\overline{TU}$

12. Refer to the rectangle below.

a. How many right angles does the rectangle have? List them.

b. Which sides are parallel?

c. Which sides are of equal length?

d. Copy the figure and draw the diagonals. Call the point where the diagonals intersect point *X*. How many diagonals does the figure have? List them.

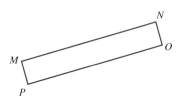

13. Fill in the blanks. In any rectangle:

a. All four angles are _____ angles.

b. Opposite sides are _____.

c. Opposite sides have equal _____.

d. The diagonals have equal _____.

e. The diagonals intersect at their _____.

14. Refer to the figure below.

a. What is m($\overline{CD}$)?   **b.** What is m($\overline{AD}$)?

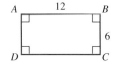

15. In the figure below, $\overline{TR}\,\|\,\overline{DF}$, $\overline{DT}\,\|\,\overline{FR}$, and m($\angle D$) = 90°. What type of quadrilateral is *DTRF*?

16. Refer to the parallelogram shown below. If m($\overline{GI}$) = 4 and m($\overline{HJ}$) = 4, what type of figure is quadrilateral *GHIJ*?

17. a. Is every rectangle a square?

b. Is every square a rectangle?

c. Is every parallelogram a rectangle?

d. Is every rectangle a parallelogram?

e. Is every rhombus a square?

f. Is every square a rhombus?

18. Trapezoid *WXYZ* is shown below. Which sides are parallel?

19. Trapezoid *JKLM* is shown below.

a. What type of trapezoid is this?

b. Which angles are the lower base angles?

c. Which angles are the upper base angles?

d. Fill in the blanks:

m($\angle J$) = _____

m($\angle K$) = _____

m($\overline{JK}$) = _____

20. Find the sum of the measures of the interior angles of the hexagon below.

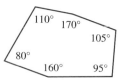

21. Refer to the figure below.

   a. How many sides does the polygon have?

   b. If you draw all of the diagonals from one vertex, how many triangles make up the polygon?

   c. What is the difference between the number of sides and the number of triangles?

   d. What formula can be used to find the sum of the interior angle measures of any polygon?

22. Refer to the regular hexagon below.

   a. Fill in the blank: ∠*BCG* is called an _____ angle of the hexagon.

   b. Fill in the blank: ∠*BCG* and _____ are adjacent and supplementary.

   c. What is m(∠*BCD*)? What is m(∠*BCG*)?

   d. What is the sum of the exterior angles of the regular hexagon?

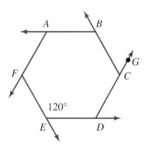

NOTATION

23. What do the tick marks in the figure indicate?

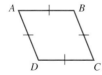

24. Rectangle *ABCD* is shown below. What do the tick marks indicate about point *X?*

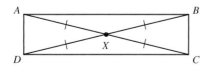

25. a. In the formula $S = (n - 2)180°$, what does S represent? What does n represent?

   b. In the formula $A = \frac{(n - 2)180°}{n}$, what does A represent? What does n represent?

   c. In the formula $E = 180° - A$, what does E represent? What does A represent?

   d. In the formula $E = \frac{360°}{n}$, what does E represent? What does n represent?

26. Suppose $n = 12$. What is $(n - 2)180°$?

GUIDED PRACTICE

In Problems 27 and 28, classify each quadrilateral as a rectangle, a square, a rhombus, or a trapezoid. Some figures may be correctly classified in more than one way. **See Objective 1.**

27. a. 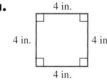   **b.**

   c.   **d.**

28. a.   **b.**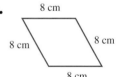

   c.   **d.**

29. Rectangle *ABCD* is shown below. **See Example 1.**

   a. What is m∠*DCB*?

   b. What is m($\overline{AX}$)?

   c. What is m($\overline{AC}$)?

   d. What is m($\overline{BD}$)?

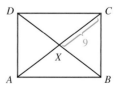

30. Refer to rectangle *EFGH* shown below. **See Example 1.**

    a. Find m∠*EHG*.

    b. Find m($\overline{FH}$).

    c. Find m($\overline{EI}$).

    d. Find m($\overline{EG}$).

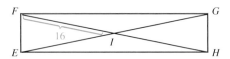

31. Refer to the trapezoid shown below. **See Example 3.**

    a. Find *x*.   **b.** Find *y*.

32. Refer to trapezoid *MNOP* shown below. **See Example 3.**

    a. Find m(∠*O*).   **b.** Find m(∠*M*).

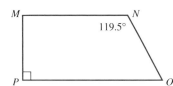

33. Refer to the isosceles trapezoid shown below. **See Example 4.**

    a. Find m($\overline{BC}$).   **b.** Find *x*.

    c. Find *y*.   **d.** Find *z*.

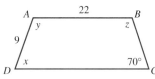

34. Refer to the trapezoid shown below. **See Example 4.**

    a. Find m(∠*T*).   **b.** Find m(∠*R*).

    c. Find m(∠*S*).

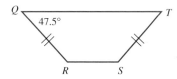

Find the sum of the interior angle measures of the polygon. **See Example 5.**

35. a 14-sided polygon   **36.** a 15-sided polygon

37. a 20-sided polygon   **38.** a 22-sided polygon

39. an octagon   **40.** a decagon

41. a dodecagon   **42.** a nonagon

Find the number of sides a polygon has if the sum of its angle measures is the given number. **See Example 6.**

43. 540°   **44.** 720°

45. 900°   **46.** 1,620°

47. 1,980°   **48.** 1,800°

49. 2,160°   **50.** 3,600°

Find the measure of one interior angle of the polygon. **See Example 7.**

51. a regular octagon   **52.** a regular nonagon

53. a regular pentagon   **54.** a regular dodecagon

55. a regular 15-sided polygon

56. a regular 20-sided polygon

57. a regular 50-sided polygon

58. a regular 100-sided polygon

Find the number of sides of a regular polygon if one of its interior angles has the following measure. **See Example 8.**

59. 120°   **60.** 90°

61. 162°   **62.** 160°

63. 150°   **64.** 156°

65. 172.8°   **66.** 179.64°

Find the measure of an exterior angle of the polygon. **See Example 9.** *(Hint: Use your answers to Problems 51–58.)*

67. a regular octagon   **68.** a regular nonagon

69. a regular pentagon   **70.** a regular dodecagon

71. a regular 15-sided polygon

72. a regular 20-sided polygon

73. a regular 50-sided polygon

74. a regular 100-sided polygon

Find the measure of each exterior angle of the polygon. **See Example 10.**

75. a regular 18-sided polygon

76. a regular 24-sided polygon

77. a regular 40-sided polygon

78. a regular 90-sided polygon

79. a regular nonagon   **80.** a regular octagon

81. a regular hexagon   **82.** a regular pentagon

TRY IT YOURSELF

83. Refer to rectangle *ABCD* shown below.

   a. Find m(∠1).

   b. Find m(∠3).

   c. Find m(∠2).

   d. If m($\overline{AC}$) is 8 cm, find m($\overline{BD}$).

   e. Find m($\overline{PD}$).

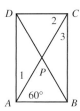

84. The following problem appeared on a quiz. Explain why the instructor must have made an error when typing the problem.

   The measure of each interior angle of a regular polygon is 169°. How many sides does the polygon have?

For Problems 85 and 86, find x. Then find the measure of each angle of the polygon.

85.

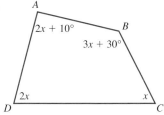

86.

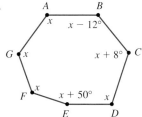

87. The measure of each interior angle of a regular polygon is $175\frac{1}{2}°$.

   a. How many sides does it have?

   b. What is the measure of an exterior angle of the polygon?

88. The measures of the exterior angles of an octagon can be represented by the algebraic expressions *x*, 2*x*, 3*x*, 4*x*, 5*x*, 6*x*, 7*x*, and 8*x*. Find the measure of each exterior angle of the octagon.

89. In the figure below, the sides of regular pentagon *ABCDE* are extended to form a five-pointed star. Find the angle measure of one of the points of the star.

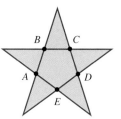

90. In the figure below, the sides of regular hexagon *ABCDEF* are extended to form a six-pointed star. Find the angle measure of one of the points of the star.

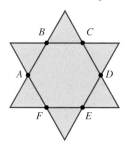

APPLICATIONS

91. QUADRILATERALS IN EVERYDAY LIFE   What quadrilateral shape do you see in each of the following objects?

   a. podium (upper portion)   **b.** checkerboard

   c. dollar bill   **d.** swimming fin

   e. camper shell window

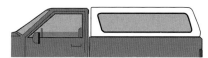

92. FLOWCHART   A flowchart shows a sequence of steps to be performed by a computer to solve a given problem. When designing a flowchart, the programmer uses a set of standardized symbols to represent various operations to be performed by the computer. Locate a rectangle, a rhombus, and a parallelogram in the flowchart shown below.

93. MAKING A FRAME   After gluing and nailing the pieces of a picture frame together, it didn't look right to a frame maker. (See the figure below.) How can she use a tape measure to make sure the corners are 90° (right) angles?

94. BASEBALL   Refer to the figure below. Find the sum of the measures of the interior angles of home plate.

95. TOOLS   The utility knife blade shown below has the shape of an isosceles trapezoid. Find *x, y,* and *z*.

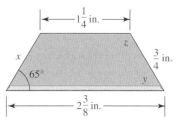

96. TRAFFIC SIGNS   The stop sign shown below is a regular polygon. What is the measure of one of its interior angles?

WRITING

97. Explain why a square is a rectangle.

98. Explain why a trapezoid is not a parallelogram.

99. Consider a regular polygon with 8 sides and the measure of one of its angles. Consider a regular polygon with 9 sides and the measure of one of its angles. Without using a formula, can you determine which interior angle measure would be larger? Explain your thinking.

100. A decagon is a polygon with ten sides. What could you call a polygon with one hundred sides? With one thousand sides? With one million sides?

6 *Perimeters and Areas of Polygons*

Objectives

1. Find the perimeter of a polygon.
2. Find the area of a polygon.
3. Find the area of figures that are combinations of polygons.

In this section, we will discuss how to find perimeters and areas of polygons. Finding perimeters is important when estimating the cost of fencing a yard or installing crown molding in a room. Finding area is important when calculating the cost of carpeting, painting a room, or fertilizing a lawn.

1. Find the perimeter of a polygon.

The **perimeter** of a polygon is the distance around it. To find the perimeter *P* of a polygon, we simply add the lengths of its sides.

Triangle

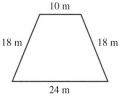

$$P = 6 + 7 + 8$$
$$= 21$$

The perimeter is 21 ft.

Quadrilateral

10 m
18 m   18 m
24 m

$$P = 10 + 18 + 24 + 18$$
$$= 70$$

The perimeter is 70 m.

Pentagon

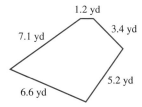

$$P = 1.2 + 7.1 + 6.6 + 5.2 + 3.4$$
$$= 23.5$$

The perimeter is 23.5 yd.

For some polygons, such as a square and a rectangle, we can simplify the computations by using a perimeter formula. Since a square has four sides of equal length *s*, its perimeter *P* is $s + s + s + s$, or $4s$.

Perimeter of a square

> If a square has a side of length *s*, its perimeter *P* is given by the formula
>
> $$P = 4s$$

EXAMPLE 1   Find the perimeter of a square whose sides are 7.5 meters long.

Strategy
We will substitute 7.5 for *s* in the formula $P = 4s$ and evaluate the right side.

Why
The variable *P* represents the unknown perimeter of the square.

Solution

$P = 4s$	This is the formula for the perimeter of a square.
$P = 4(\mathbf{7.5})$	Substitute 7.5 for *s*, the length of one side of the square.
$P = 30$	Do the multiplication.

The perimeter of the square is 30 meters.

Self Check 1
A Scrabble game board has a square shape with sides of length 38.5 cm. Find the perimeter of the game board.

Answer: 154 cm

Now Try
Problems 17 and 19   ▪

Since a rectangle has two lengths *l* and two widths *w*, its perimeter *P* is *l* + *w* + *l* + *w*, or 2*l* + 2*w*.

Perimeter of a rectangle

If a rectangle has length *l* and width *w*, its perimeter *P* is given by the formula

$$P = 2l + 2w$$

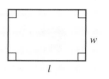

 CAUTION   When finding the perimeter of a polygon, the lengths of the sides must be expressed in the same units.

EXAMPLE 2   Find the perimeter of the rectangle shown on the right, in meters.

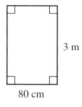

Strategy
We will express the length of the rectangle in meters and then use the formula *P* = 2*l* + 2*w* to find the perimeter of the figure.

Why
We can only add quantities that are measured in the same units.

Solution
Since 1 meter = 100 centimeters, we can convert 80 centimeters to meters by multiplying 80 centimeters by the unit conversion factor $\frac{1 \text{ m}}{100 \text{ cm}}$.

$$80 \text{ cm} = 80 \text{ cm} \cdot \frac{1 \text{ m}}{100 \text{ cm}} \qquad \text{Multiply by 1: } \tfrac{1 \text{ m}}{100 \text{ cm}} = 1.$$

$$= \frac{80 \cancel{\text{ cm}}^{\,1}}{1} \cdot \frac{1 \text{ m}}{100 \cancel{\text{ cm}}_{\,1}} \qquad \begin{array}{l}\text{Write 80 cm as a fraction: } 80 \text{ cm} = \tfrac{80 \text{ cm}}{1}. \text{ Remove the} \\ \text{common units of centimeters in the numerator and} \\ \text{denominator.}\end{array}$$

$$= \frac{80}{100} \text{ m}$$

$$= 0.8 \text{ m} \qquad \begin{array}{l}\text{Divide by 100 by moving the understood decimal point} \\ \text{in 80 two places to the left.}\end{array}$$

The length of the rectangle is 0.8 m. We can now substitute 0.8 for *l*, the length, and 3 for *w*, the width, in the formula for the perimeter of a rectangle.

$$P = 2l + 2w \qquad \text{This is the formula for the perimeter of a rectangle.}$$

$$P = 2(\mathbf{0.8}) + 2(\mathbf{3}) \quad \text{Substitute for } l \text{ and } w.$$

$$= 1.6 + 6 \qquad \text{Do the multiplication.}$$

$$= 7.6 \qquad \text{Do the addition.}$$

The perimeter of the rectangle is 7.6 meters.

Self Check 2
Find the perimeter of the triangle shown below, in inches.

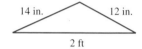

Answer: 50 in.

Now Try
Problem 21

EXAMPLE 3   *Structural engineering.*   The truss shown below is made up of three parts that form an isosceles triangle. The length of the base is 4 feet less than twice the length of one of the sides. If 76 linear feet of lumber were used to make the truss, how long is each part of the truss?

Analyze the problem
• The truss is in the shape of an isosceles triangle.
• The length of the base is 4 feet less than twice the length of a side.
• The perimeter of the truss is 76 feet.
• We are to find the length of each part of the truss.

Form an equation

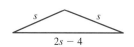

Since the length of the base is related to the length of a side, we begin by letting s = the length of a side of the truss (in feet). Then we translate the words *4 feet less than twice the length of one side,* to get an algebraic expression to represent the length of the base.

$$2s - 4 = \text{the length of the base (in feet)}$$

At this stage, it is helpful to draw a sketch. (See the figure on the left.)

Because 76 linear feet of lumber were used to make the triangular-shaped truss,

The length of the base of the truss	plus	the length of one side	plus	the length of the other side	equals	the perimeter of the truss.
$2s - 4$	$+$	s	$+$	s	$=$	76

Solve the equation

$$2s - 4 + s + s = 76$$

$4s - 4 = 76$ Combine like terms: $2s + s + s = 4s$.

$4s = 80$ To isolate the variable term, $4s$, undo the subtraction of 4 by adding 4 to both sides.

$\dfrac{4s}{4} = \dfrac{80}{4}$ To isolate the variable s, undo the multiplication by 4 by dividing both sides by 4.

$s = 20$ Do the divisions.

To find the length of the base, we evaluate the expression $2s - 4$ for $s = 20$.

$2s - 4 = 2(\mathbf{20}) - 4$ Substitute 20 for s.

$ = 36$

State the conclusion   The length of the base of the truss is 36 ft, and the length of each side is 20 ft.

Check the result   If we add the lengths of the parts of the truss, we get 36 ft + 20 ft + 20 ft = 76 ft. The answers check.

Now Try
Problem 25

■

Using Your Calculator: **Perimeters of figures that are combinations of polygons**

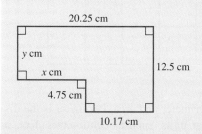

To find the perimeter of the figure shown on the left, we need to know the values of x and y. Since the figure is a combination of two rectangles, we can use a calculator to see that

$x = 20.25 - 10.17$ and $y = 12.5 - 4.75$

$x = 10.08$ cm $y = 7.75$ cm

The perimeter P of the figure is

$$P = 20.25 + 12.5 + 10.17 + 4.75 + x + y$$
$$P = 20.25 + 12.5 + 10.17 + 4.75 + \mathbf{10.08} + 7.75$$

We can use a scientific calculator to make this calculation.

Keystrokes

The perimeter is 65.5 centimeters.

2. Find the area of a polygon.

The **area** of a polygon is the measure of the amount of surface it encloses. Area is measured in square units, such as square inches or square centimeters, as shown below.

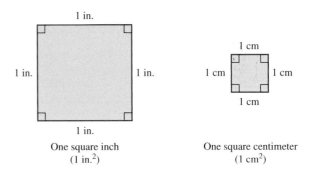

One square inch
(1 in.^{2})

One square centimeter
(1 cm^2)

In everyday life, we often use areas. For example,

- To carpet a room, we buy square yards.
- A can of paint will cover a certain number of square feet.
- To measure vast amounts of land, we often use square miles.
- We buy house roofing by the "square." One square is 100 square feet.

The rectangle shown below has a length of 10 centimeters and a width of 3 centimeters. If we divide the rectangle into squares as shown in the figure, each square represents an area of 1 square centimeter—a surface enclosed by a square measuring 1 centimeter on each side. Because there are 3 rows with 10 squares in each row, there are 30 squares. Since the rectangle encloses a surface area of 30 squares, its area is 30 square centimeters, which can be written as 30 cm^2.

This example illustrates that to find the area of a rectangle, we multiply its length by its width.

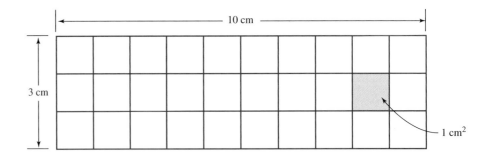

 CAUTION Do not confuse the concepts of perimeter and area. Perimeter is the distance around a polygon. It is measured in linear units, such as centimeters, feet, or miles. Area is a measure of the surface enclosed within a polygon. It is measured in square units, such as square centimeters, square feet, or square miles.

In practice, we do not find areas of polygons by counting squares. Instead, we use formulas to find areas of geometric figures. We have seen that the area of a rectangle is the product of its length and width. This fact can be used to derive the area formula for a parallelogram.

The figure on the next page shows how a parallelogram with base b and height h can be transformed into a rectangle with length b and width h. Since the area of the rectangle is bh, the area of the parallelogram is also *bh*. Therefore, the area A of the parallelogram is given by the formula $A = bh$.

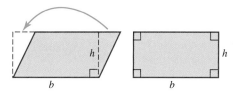

To derive the formula for the area of a triangle, we use the fact that the area of a parallelogram is the product of the length of its base and its height. The figure below shows how a triangle with base *b* and height *h* can, with the addition of an identical triangle, be transformed into a parallelogram with base *b* and height *h*. Since the area of the parallelogram is *bh*, the area of the original triangle must be one-half of that. Therefore, the area A of the original triangle is given by the formula $A = \frac{1}{2}bh$.

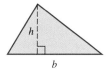 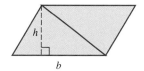

We can derive the formula for a trapezoid using the formula for the area of a parallelogram. The following figure shows how a trapezoid with bases b_1 and b_2 and height *h* can, with the addition of an identical trapezoid, be transformed into a parallelogram with base $(b_1 + b_2)$ and height *h*. Since the area of the parallelogram is $(b_1 + b_2)h$, the area of the original trapezoid must be one-half of that. Therefore, the area A of the original trapezoid is given by the formula $A = \frac{1}{2}(b_1 + b_2)h$.

The factors on the right side of the formula for the area of a trapezoid are usually written in a different order. By the commutative property of multiplication, we have:

$$A = \frac{1}{2}h(b_1 + b_2).$$

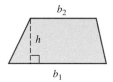 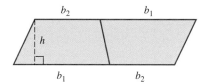

The formulas for finding the area of several types of polygons are summarized in the table on the next page.

EXAMPLE 4   Find the area of the square shown on the right.

Strategy
We will substitute 15 for *s* in the formula $A = s^2$ and evaluate the right.

Why
The variable *A* represents the unknown area of the square.

Solution

$A = s^2$ This is the formula for the area of a square.

$A = 15^2$ Substitute 15 for *s*, the length of one side of the square.

$A = 225$ Evaluate the exponential expression: $15 \cdot 15 = 225$.

Recall that area is measured in square units. Thus, the area of the square is 225 square centimeters, which can be written as 225 cm^2.

Self Check 4
Find the area of the square shown below.

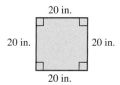

Answer: 400 in.^{2}

Now Try
Problems 29 and 31

Figure	**Name**	**Formula for area**
	Square	$A = s^2$, where s is the length of one side.
	Rectangle	$A = lw$, where l is the length and w is the width.
	Parallelogram	$A = bh$, where b is the length of the base and h is the height. (A height is always perpendicular to the base.)
	Triangle	$A = \frac{1}{2}bh$, where b is the length of the base and h is the height. The segment perpendicular to the base and representing the height is called an **altitude**.
	Trapezoid	$A = \frac{1}{2}h(b_1 + b_2)$, where h is the height of the trapezoid and b_1 and b_2 represent the lengths of the bases.

EXAMPLE 5   Find the number of square feet in 1 square yard.

Strategy

A figure is helpful to solve this problem. We will draw a square yard and divide each of its sides into 3 equally long parts.

Why

Since a square yard is a square with each side measuring 1 yard, each side also measures 3 feet.

Solution

$$1 \text{ yd}^2 = (\mathbf{1 \text{ yd}})^2$$
$$= (\mathbf{3 \text{ ft}})^2 \quad \text{Substitute 3 feet for 1 yard.}$$
$$= (3 \text{ ft})(3 \text{ ft})$$
$$= 9 \text{ ft}^2$$

There are 9 square feet in 1 square yard.

Self Check 5

Find the number of square centimeters in 1 square meter.

Answer: $10,000 \text{ cm}^2$

Now Try

Problems 33 and 39

EXAMPLE 6   *Women's sports.*

Field hockey is a team sport in which players use sticks to try to hit a ball into their opponents' goal. Find the area of the rectangular field shown on the right. Give the answer in square feet.

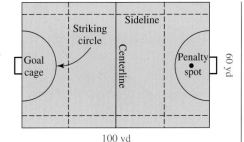

100 yd

Strategy
We will substitute 100 for *l* and 50 for *w* in the formula $A = lw$ and evaluate the right side.

Why
The variable *A* represents the unknown area of the rectangle.

Solution

$A = lw$ This is the formula for the area of a rectangle.

$A = 100(60)$ Substitute 100 for *l*, the length, and 60 for *w*, the width.

$ = 6{,}000$ Do the multiplication.

The area of the rectangle is 6,000 square yards. Since there are 9 square feet per square yard, we can convert this number to square feet by multiplying 6,000 square yards by $\frac{9 \text{ ft}^2}{1 \text{ yd}^2}$.

$6{,}000 \text{ yd}^2 = 6{,}000 \text{ yd}^2 \cdot \dfrac{9 \text{ ft}^2}{1 \text{ yd}^2}$ Multiply by the unit conversion factor: $\frac{9 \text{ ft}^2}{1 \text{ yd}^2}$.

$\phantom{6{,}000 \text{ yd}^2} = 6{,}000 \cdot 9 \text{ ft}^2$ Remove the common units of square yards in the numerator and denominator.

$\phantom{6{,}000 \text{ yd}^2} = 54{,}000 \text{ ft}^2$ Multiply: $6{,}000 \cdot 9 = 54{,}000$.

The area of the field is 54,000 ft^2.

Self Check 6
A regulation-size ping-pong table is 9 feet long and 5 feet wide. Find its area in square inches.

Answer: 6,480 in.^{2}

Now Try
Problem 41

EXAMPLE 7   Find the area of the triangle shown on the right.

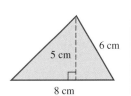

Strategy
We will substitute 8 for *b* and 5 for *h* in the formula $A = \frac{1}{2}bh$ and evaluate the right side. (The side having length 6 cm is additional information that is not used to find the area.)

Why
The variable *A* represents the unknown area of the triangle.

Solution

$A = \frac{1}{2}bh$ This is the formula for the area of a triangle.

$A = \frac{1}{2}(8)(5)$ Substitute 8 for *b*, the length of the base, and 5 for *h*, the height.

$ = 4(5)$ Do the first multiplication: $\frac{1}{2}(8) = 4$.

$ = 20$ Complete the multiplication.

The area of the triangle is 20 cm^2.

Self Check 7
Find the area of the triangle shown below.

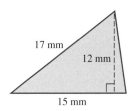

Answer: 90 mm^2

Now Try
Problem 45

EXAMPLE 8   Find the area of the triangle shown on the right.

Strategy
We will substitute 9 for b and 13 for h in the formula $A = \frac{1}{2}bh$ and evaluate the right side.

Why
The variable A represents the unknown area of the triangle.

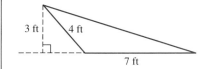

Solution
In this case, the altitude falls outside the triangle.

$A = \frac{1}{2}bh$ This is the formula for the area of a triangle.

$A = \frac{1}{2}(9)(13)$ Substitute 9 for b, the length of the base, and 13 for h, the height.

$\quad = \frac{1}{2}\left(\frac{9}{1}\right)\left(\frac{13}{1}\right)$ Write 9 as $\frac{9}{1}$ and 13 as $\frac{13}{1}$.

$\quad = \frac{117}{2}$ Multiply the fractions.

$\quad = 58.5$ Do the division.

The area of the triangle is 58.5 cm^2.

Self Check 8
Find the area of the triangle shown below.

Answer: 10.5 ft^2

Now Try
Problem 49

EXAMPLE 9   Find the area of the trapezoid shown on the right.

Strategy
We will express the height of the trapezoid in inches and then use the formula $A = \frac{1}{2}h(b_1 + b_2)$ to find the area of the figure.

Why
The height of 1 foot must be expressed as 12 inches to be consistent with the units of the bases.

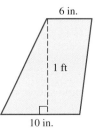

Solution

$A = \frac{1}{2}h(b_1 + b_2)$ This is the formula for the area of a trapezoid.

$A = \frac{1}{2}(12)(10 + 6)$ Substitute 12 for h, the height; 10 for b_1, the length of the lower base; and 6 for b_2, the length of the upper base.

$\quad = \frac{1}{2}(12)(16)$ Do the addition within the parentheses.

$\quad = 6(16)$ Do the first multiplication: $\frac{1}{2}(12) = 6$.

$\quad = 96$ Complete the multiplication

The area of the trapezoid is 96 in^2.

Self Check 9
Find the area of the trapezoid shown below.

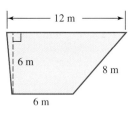

Answer: 54 m^2

Now Try
Problem 53

EXAMPLE 10   The area of the parallelogram shown on the right is 360 ft². Find the height.

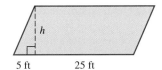

Strategy
To find the height of the parallelogram, we will substitute the given values in the formula $A = bh$ and solve for h.

Why
The variable h represents the unknown height.

Solution
From the figure, we see that the length of the base of the parallelogram is

5 feet + 25 feet = 30 feet

$A = bh$ This is the formula for the area of a parallelogram.

$360 = 30h$ Substitute 360 for A, the area, and 30 for b, the length of the base.

$\dfrac{360}{30} = \dfrac{30h}{30}$ To isolate h, undo the multiplication by 30 by dividing both sides by 30.

$12 = h$ Do the divisions.

The height of the parallelogram is 12 feet.

Self Check 10
The area of the parallelogram below is 96 cm². Find its height.

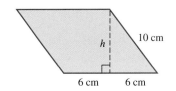

Answer: 8 cm

Now Try
Problems 57 and 61

---

3. Find the area of figures that are combinations of polygons.

EXAMPLE 11   Find the area of one side of the tent shown on the right.

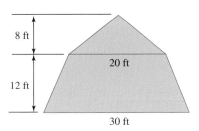

Strategy
We will use the formula $A = \frac{1}{2}h(b_1 + b_2)$ to find the area of the lower portion of the tent and the formula $A = \frac{1}{2}bh$ to find the area of the upper portion of the tent. Then we will combine the results.

Why
A side of the tent is a combination of a trapezoid and a triangle.

Solution
To find the area of the lower portion of the tent, we proceed as follows.

$A_{\text{trap.}} = \frac{1}{2}h(b_1 + b_2)$ This is the formula for the area of a trapezoid.

$A_{\text{trap.}} = \frac{1}{2}(12)(30 + 20)$ Substitute 30 for b_1, 20 for b_2, and 12 for h.

$= \frac{1}{2}(12)(50)$ Do the addition within the parentheses.

$= 6(50)$ Do the first multiplication: $\frac{1}{2}(12) = 6$.

$= 300$ Complete the multiplication.

The area of the trapezoid is 300 ft².

Self Check 11
Find the area of the shaded figure below.

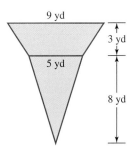

To find the area of the upper portion of the tent, we proceed as follows.

$A_{\text{triangle}} = \frac{1}{2}bh$ This is the formula for the area of a triangle.

$A_{\text{triangle}} = \frac{1}{2}(20)(8)$ Substitute 20 for b and 8 for h.

$\qquad = 80$ Do the multiplications, working from left to right: $\frac{1}{2}(20) = 10$ and then $10(8) = 80$.

The area of the triangle is 80 ft^2.

The total area of one side of the tent is

$A_{\text{total}} = A_{\text{trap.}} + A_{\text{triangle}}$

$A_{\text{total}} = 300 \text{ ft}^2 + 80 \text{ ft}^2$

$\qquad = 380 \text{ ft}^2$

The total area is 380 ft^2.

Answer: 41 yd^2

Now Try
Problem 65

EXAMPLE 12   Find the area of the shaded region shown on the right.

Strategy
We will subtract the unwanted area of the square from the area of the rectangle.

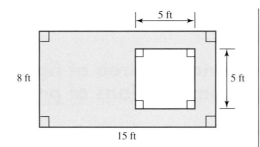

Self Check 12
Find the area of the shaded region shown below.

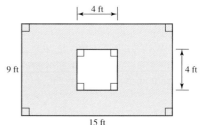

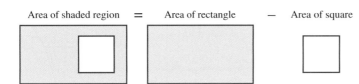

Area of shaded region = Area of rectangle — Area of square

Why
The area of the rectangular-shaped shaded figure does not include the square region inside of it.

Solution

$A_{\text{shaded}} = lw - s^2$ The formula for the area of a rectangle is $A = lw$, and the formula for the area of a square is $A = s^2$.

$\qquad = 15(8) - 5^2$ To find the area of the rectangle, substitute 15 for the length l and 8 for the width w. To find the area of the square, substitute 5 for the length s of a side.

$\qquad = 120 - 25$

$\qquad = 95$

The area of the shaded region is 95 ft^2.

Answer: 119 ft^2

Now Try
Problem 69

EXAMPLE 13 *Carpeting a room.* A living room/dining room has the floor plan shown in the figure. If carpet costs $29 per square yard, including pad and installation, how much will it cost to carpet the room? (Assume no waste.)

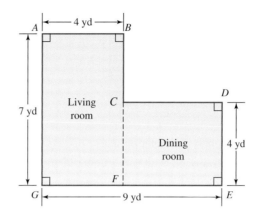

Strategy
We will find the number of square yards of carpeting needed and multiply the result by $29.

Why
Each square yard costs $29.

Solution
First, we must find the total area of the living room and the dining room:

$$A_{\text{total}} = A_{\text{living room}} + A_{\text{dining room}}$$

Since $\overline{CF}$ divides the space into two rectangles, the areas of the living room and the dining room are found by multiplying their respective lengths and widths. Therefore, the area of the living room is 4 yd · 7 yd = 28 yd^2.

The width of the dining room is given as 4 yd. To find its length, we subtract:

$$\text{m}(\overline{CD}) = \text{m}(\overline{GE}) - \text{m}(\overline{AB}) = 9 \text{ yd} - 4 \text{ yd} = 5 \text{ yd}$$

Thus, the area of the dining room is 5 yd · 4 yd = 20 yd^2. The total area to be carpeted is the sum of these two areas.

$$A_{\text{total}} = A_{\textbf{living room}} + A_{\text{dining room}}$$
$$A_{\text{total}} = \textbf{28 yd}^2 + 20 \text{ yd}^2$$
$$= 48 \text{ yd}^2$$

At $29 per square yard, the cost to carpet the room will be 48 · $29, or $1,392.

Now Try
Problem 73

STUDY SET Section 6

VOCABULARY *Fill in the blanks.*

1. The distance around a polygon is called the

_____.

2. The _____ of a polygon is measured in linear units such as inches, feet, and miles.

3. The measure of the surface enclosed by a polygon is called its _____.

4. If each side of a square measures 1 foot, the area enclosed by the square is 1 _____ foot.

5. The _____ of a polygon is measured in square units.

6. The segment that represents the height of a triangle is called an _____.

CONCEPTS

7. The figure below shows a kitchen floor that is covered with 1-foot-square tiles. What is the area of the floor?

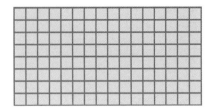

8. Tell which concept applies, perimeter or area.
   a. The length of a walk around New York's Central Park
   b. The amount of office floor space in the White House
   c. The amount of fence needed to enclose a playground
   d. The amount of land in Yellowstone National Park

9. Give the formula for the perimeter of a
   a. square   **b.** rectangle

10. Give the formula for the area of a
   a. square
   b. rectangle
   c. triangle
   d. trapezoid
   e. parallelogram

11. For each figure below, draw the altitude to the base *b*.

   a.   **b.**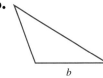

   c. 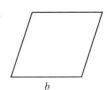   **d.**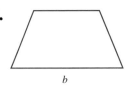

12. For each figure below, label the base *b* for the given altitude.

   a. 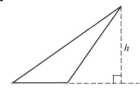   **b.**

   c.   **d.**

13. The shaded figure below is a combination of what two types of geometric figures?

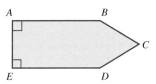

14. Explain how you would find the area of the following shaded figure.

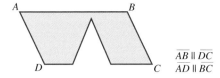

$\overline{AB} \parallel \overline{DC}$
$\overline{AD} \parallel \overline{BC}$

NOTATION   *Fill in the blanks.*

15. a. The symbol 1 in.^{2} means one _____.
   b. One square meter is expressed as _____.

16. In the figure below, the symbol ⌐ indicates that the dashed line segment, called an *altitude*, is _____ to the base.

GUIDED PRACTICE   *Find the perimeter of each square. **See Example 1.***

17.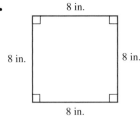

18.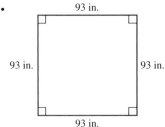

19. A square with sides 5.75 miles long

20. A square with sides 3.4 yards long

Find the perimeter of each rectangle, in meters.
See Example 2.

21.

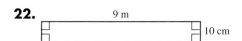

12 m
60 cm

22.

9 m
10 cm

23.   90 cm   **24.**   70 cm

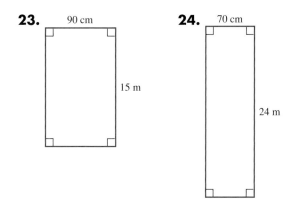

15 m

24 m

Write and then solve an equation to answer each problem. **See Example 3.**

25. The perimeter of an isosceles triangle is 35 feet. The base is 5 feet less than twice as long as one of the congruent sides. Find the length of each side of the triangle.

26. The perimeter of an isosceles triangle is 94 feet. Each of the congruent sides is 2 feet more than four times as long as the base. Find the length of each side of the triangle.

27. The perimeter of an isosceles trapezoid is 35 meters. The upper base is 5 meters shorter than the lower base. Each leg is 10 meters shorter than the lower base. How long is each side of the trapezoid?

28. The perimeter of an isosceles trapezoid is 46 inches. The upper base is 3 inches longer than one of the congruent legs, and the lower base is 2 inches less than twice as long as one of the congruent legs. How long is each side of the trapezoid?

Find the area of each square. **See Example 4.**

29.

4 cm

4 cm

30.

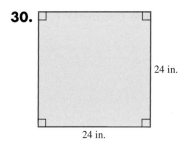

24 in.

24 in.

31. A square with sides 2.5 meters long
32. A square with sides 6.8 feet long

For Problems 33–40, **see Example 5.**

33. How many square inches are in 1 square foot?
34. How many square inches are in 1 square yard?
35. How many square millimeters are in 1 square meter?
36. How many square decimeters are in 1 square meter?
37. How many square feet are in 1 square mile?
38. How many square yards are in 1 square mile?
39. How many square meters are in 1 square kilometer?
40. How many square decameters are in 1 square kilometer?

Find the area of each rectangle. Give the answer in square feet. **See Example 6.**

41.

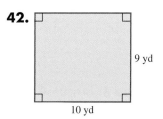

3 yd

5 yd

42.

9 yd

10 yd

43.

20 yd

62 yd

44.

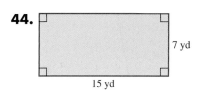

7 yd

15 yd

Find the area of each triangle. **See Example 7.**

45.

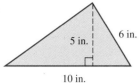

46.

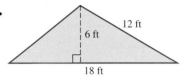

47.

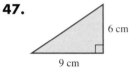

48.

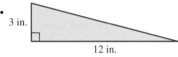

Find the area of each triangle. **See Example 8.**

49.　　　　**50.**

51.

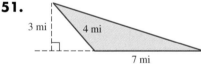

52.

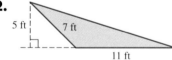

Find the area of each trapezoid. **See Example 9.**

53.

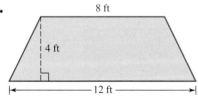

54.

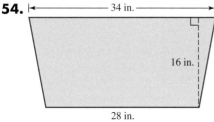

55.

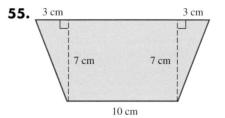

56.

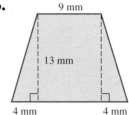

Solve each problem. **See Example 10.**

57. The area of a parallelogram is 60 m², and its height is 15 m. Find the length of its base.

58. The area of a parallelogram is 95 in.², and its height is 5 in. Find the length of its base.

59. The area of a rectangle is 36 cm², and its length is 3 cm. Find its width.

60. The area of a rectangle is 144 mi², and its length is 6 mi. Find its width.

61. The area of a triangle is 54 m², and the length of its base is 3 m. Find the height.

62. The area of a triangle is 270 ft², and the length of its base is 18 ft. Find the height.

63. The perimeter of a rectangle is 64 mi, and its length is 21 mi. Find its width.

64. The perimeter of a rectangle is 26 yd, and its length is 10.5 yd. Find its width.

Find the area of each shaded figure. **See Example 11.**

65.

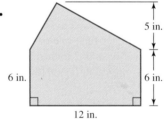

66.

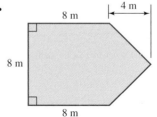

67.

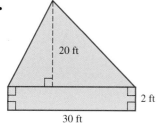

68.

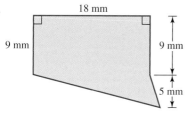

Find the area of each shaded figure. **See Example 12.**

69.

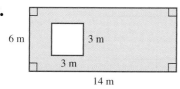

70.

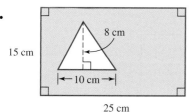

71.

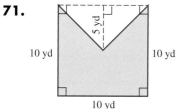

72.

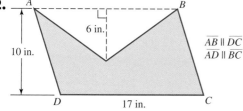

$\overline{AB} \parallel \overline{DC}$
$\overline{AD} \parallel \overline{BC}$

Solve each problem. **See Example 13.**

73. FLOORING   A rectangular family room is 8 yards long and 5 yards wide. At $30 per square yard, how much will it cost to put down vinyl sheet flooring in the room? (Assume no waste.)

74. CARPETING   A rectangular living room measures 10 yards by 6 yards. At $32 per square yard, how much will it cost to carpet the room? (Assume no waste.)

75. FENCES   A man wants to enclose a rectangular yard with fencing that costs $12.50 a foot, including installation. Find the cost of enclosing the yard if its dimensions are 110 ft by 85 ft.

76. FRAMES   Find the cost of framing a rectangular picture with dimensions of 24 inches by 30 inches if framing material costs $0.75 per inch.

TRY IT YOURSELF

Sketch and label each of the figures.

77. Two different rectangles, each having a perimeter of 40 in.

78. Two different rectangles, each having an area of 40 in.^{2}

79. A square with an area of 25 m^2

80. A square with a perimeter of 20 m

81. A parallelogram with an area of 15 yd^2

82. A triangle with an area of 20 ft^2

83. A figure consisting of a combination of two rectangles, whose total area is 80 ft^2

84. A figure consisting of a combination of a rectangle and a square, whose total area is 164 ft^2

Find the area of each parallelogram.

85.

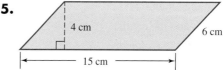

86.

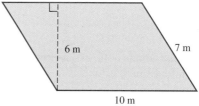

87. The lengths of the sides of the polygons are represented by algebraic expressions. In each case, the units are feet. Write an algebraic expression that represents the perimeter of the polygon.

a.

b.

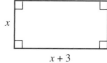

c.

d.

88. The dimensions of the polygons below are represented by algebraic expressions. In each case, the units are meters. Write an algebraic expression that represents the area of the polygon.

a. 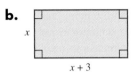   **b.**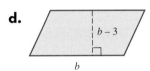

c. 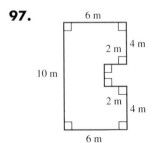   **d.**

89. The perimeter of an isosceles triangle is 80 meters. If the length of one of the congruent sides is 22 meters, how long is the base?

90. The perimeter of a square is 35 yards. How long is a side of the square?

91. The perimeter of an equilateral triangle is 85 feet. Find the length of each side.

92. An isosceles triangle with congruent sides of length 49.3 inches has a perimeter of 121.7 inches. Find the length of the base.

93. The perimeter of a rectangle is 60 cm, and the width is 12 cm less than the length.

   a. Find its width and its length.

   b. Find the area of the rectangle.

94. The perimeter of a rectangle is 38 m, and the length is 1 m more than the width.

   a. Find its width and its length.

   b. Find the area of the rectangle.

95. The height of a trapezoid is 6 in., and its lower base is 1 inch more than four times as long as its upper base. If the area of the trapezoid is 48 in.^{2}, find the length of the upper base and the lower base.

96. The height of a trapezoid is 10 cm, and its lower base is 2 cm less than three times as long as its upper base. If the area of the trapezoid is 130 cm^2, find the length of the upper base and the lower base.

Find the perimeter of the figure.

97.

98.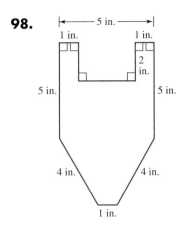

Find x and y. Then find the perimeter of the figure.

99.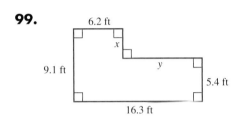

100.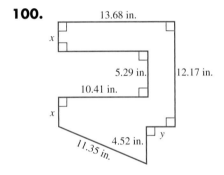

APPLICATIONS

101. LANDSCAPING A woman wants to plant a pine-tree screen around three sides of her backyard. (See the figure below.) If she plants the trees 3 feet apart, how many trees will she need?

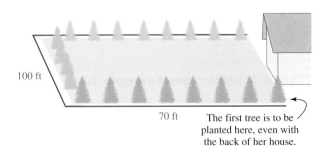

100 ft

70 ft   The first tree is to be planted here, even with the back of her house.

102. GARDENING   A gardener wants to plant a border of marigolds around the garden shown below, to keep out rabbits. How many plants will she need if she allows 6 inches between plants?

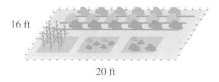

16 ft

20 ft

103. COMPARISON SHOPPING   Which is more expensive: a ceramic-tile floor costing $3.75 per square foot or linoleum costing $34.95 per square yard?

104. COMPARISON SHOPPING   Which is cheaper: a hardwood floor costing $6.95 per square foot or a carpeted floor costing $37.50 per square yard?

105. TILES   A rectangular basement room measures 14 by 20 feet. Vinyl floor tiles that are 1 ft^2 cost $1.29 each. How much will the tile cost to cover the floor? (Assume no waste.)

106. PAINTING   The north wall of a barn is a rectangle 23 feet high and 72 feet long. There are five windows in the wall, each 4 by 6 feet. If a gallon of paint will cover 300 ft^2, how many gallons of paint must the painter buy to paint the wall?

107. SAILS   If nylon is $12 per square yard, how much would the fabric cost to make a triangular sail with a base of 12 feet and a height of 24 feet?

108. REMODELING   The gable end of a house is an isosceles triangle with a height of 4 yards and a base of 23 yards. It will require one coat of primer and one coat of finish to paint the triangle. Primer costs $17 per gallon, and the finish paint costs $23 per gallon. If one gallon of each type of paint covers 300 square feet, how much will it cost to paint the gable, excluding labor?

109. GEOGRAPHY   Use the dimensions of the trapezoid that is superimposed over the state of Nevada to estimate the area of the "Silver State."

OREGON

IDAHO

315 mi

205 mi

NEVADA

● Reno

★ Carson City

505 mi

CALIFORNIA

UTAH

Las Vegas ●

ARIZONA

110. SOLAR COVERS   A swimming pool has the shape shown below. How many square feet of a solar blanket material will be needed to cover the pool? How much will the cover cost if it is $1.95 per square foot? (Assume no waste.)

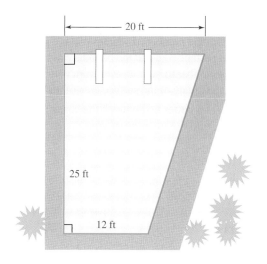

20 ft

25 ft

12 ft

111. CARPENTRY   How many sheets of 4-foot-by-8-foot sheetrock are needed to drywall the inside walls on the first floor of the barn shown below? (Assume that the carpenters will cover each wall entirely and then cut out areas for the doors and windows.)

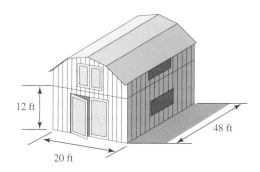

12 ft

48 ft

20 ft

112. CARPENTRY   If it costs $90 per square foot to build a one-story home in northern Wisconsin, find the cost of building the house with the floor plan shown below.

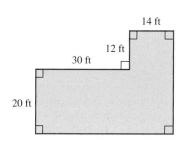

14 ft

12 ft

30 ft

20 ft

113. ESTIMATING SURFACE AREA   In figure (a) below, a grid is superimposed over a picture of a lake. Each square is 1 mile on a side and therefore represents 1 square mile.

a. Count the number of squares that are completely within the boundary (shoreline) of the lake. This is an underestimate of the surface area of the lake.

b. Count the number of squares that are partially inside and partially outside the boundary of the lake. Add this number to your answer from part a. This is an overestimate of the surface area of the lake.

c. To get a better estimate of the surface area of the lake, find the *average* of your answers to parts a and b.

d. In figure (b), a grid of squares with sides of length $\frac{1}{2}$ mile is superimposed over the same picture of the lake. Repeat the above process, but keep in mind that each square covers only $\frac{1}{4}$ of a square mile.

(a)

(b)

114. ELECTRIC IRONS   See the illustration below. Estimate the area of the sole plate of the iron by thinking of it as a combination of a trapezoid and a triangle.

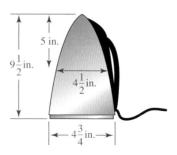

WRITING

115. Explain the difference between perimeter and area.

116. Why is it necessary that area be measured in square units?

117. A student expressed the area of the square in the figure below as 25^2 ft. Explain his error.

5 ft

5 ft

118. Refer to the figure below. What must be done before we can use the formula to find the area of this rectangle?

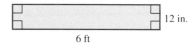

12 in.

6 ft

7 *Circles*

Objectives

1. Define circle, radius, chord, diameter, and arc.
2. Find the circumference of a circle.
3. Find the area of a circle.
4. Find the length of a circular arc.
5. Find the area of a sector.

In this section, we will discuss the circle, one of the most useful geometric figures of all. In fact, the discoveries of fire and the circular wheel are two of the most important events in the history of the human race. We will begin our study by introducing some basic vocabulary associated with circles.

1. Define circle, radius, chord, diameter, and arc.

Circle

> A **circle** is the set of all points in a plane that lie a fixed distance from a point called its **center.**

A segment drawn from the center of a circle to a point on the circle is called a **radius.** (The plural of *radius* is *radii.*) From the definition, it follows that all radii of the same circle are the same length.

A **chord** of a circle is a line segment that connects two points on the circle. A **diameter** is a chord that passes through the center of the circle. Since a diameter D of a circle is twice as long as a radius r, we have

$$D = 2r$$

Each of the previous definitions is illustrated in figure (a) below, in which O is the center of the circle.

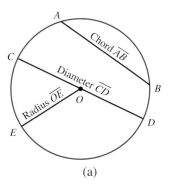
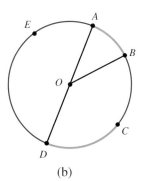

(a)   (b)

Any part of a circle is called an **arc.** In figure (b) above, the part of the circle from point A to point B is $\overset{\frown}{AB}$, read as "arc AB." $\overset{\frown}{CD}$ is the part of the circle from point C to point D. An arc that is half of a circle is a **semicircle.**

Semicircle

> A **semicircle** is an arc of a circle whose endpoints are the endpoints of a diameter.

If point O is the center of the circle in figure (b), $\overline{AD}$ is a diameter and $\overset{\frown}{AED}$ is a semicircle. The middle letter E distinguishes semicircle $\overset{\frown}{AED}$ (the part of the circle from point A to point D that includes point E) from semicircle $\overset{\frown}{ABD}$ (the part of the circle from point A to point D that includes point B).

An arc that is shorter than a semicircle is a **minor arc.** An arc that is longer than a semicircle is a **major arc.** In figure (b),

$\overset{\frown}{AE}$ is a minor arc   and $\overset{\frown}{ABE}$ is a major arc.

 CAUTION It is often possible to name a major arc in more than one way. For example, in figure (b), major arc $\overset{\frown}{ABE}$ is the part of the circle from point A to point E that includes point B. Two other names for the same major arc are $\overset{\frown}{ACE}$ and $\overset{\frown}{ADE}$.

2. Find the circumference of a circle.

Since early history, mathematicians have known that the ratio of the distance around a circle (the circumference) divided by the length of its diameter is approximately 3. First Kings, Chapter 7, of the Bible describes a round bronze tank that was 15 feet from brim to brim and 45 feet in circumference, and $\frac{45}{15} = 3$. Today, we have a better value for this ratio, known as π (pi). If C is the circumference of a circle and D is the length of its diameter, then

$$\pi = \frac{C}{D} \qquad \text{where } \pi = 3.141592653589\ldots \quad \tfrac{22}{7} \text{ and } 3.14 \text{ are often used as estimates of } \pi.$$

If we multiply both sides of $\pi = \frac{C}{D}$ by D, we have the following formula.

Circumference of a circle

The circumference of a circle is given by the formula

$C = \pi D$ where C is the circumference and D is the length of the diameter

Since a diameter of a circle is twice as long as a radius r, we can substitute $2r$ for D in the formula $C = \pi D$ to obtain another formula for the circumference C:

$C = 2\pi r$ The notation $2\pi r$ means $2 \cdot \pi \cdot r$.

EXAMPLE 1   Find the circumference of the circle shown on the right. Give the exact answer and an approximation.

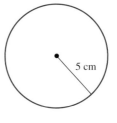

Strategy
We will substitute 5 for r in the formula $C = 2\pi r$ and evaluate the right side.

Why
The variable C represents the unknown circumference of the circle.

Solution

$C = 2\pi r$ This is the formula for the circumference of a circle.

$C = 2\pi(5)$ Substitute 5 for r, the radius.

$C = 2(5)\pi$ Normally, when a product involves π, we rewrite it so that π is the last factor.

$C = 10\pi$ Do the first multiplication: $2(5) = 10$. This is the exact answer.

The circumference of the circle is exactly 10π cm. If we replace π with 3.14, we get an approximation of the circumference.

$C = 10\pi$

$C \approx 10(\mathbf{3.14})$

$C \approx 31.4$ To multiply by 10, move the decimal point in 3.14 one place to the right.

The circumference of the circle is approximately 31.4 cm.

Self Check 1
Find the circumference of the circle shown below. Give the exact answer and an approximation.

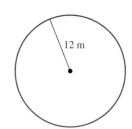

Answer: 24π m ≈ 75.4 m

Now Try
Problem 45

Using Your Calculator: *Calculating revolutions of a tire*

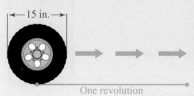

←15 in.→

One revolution

When the $\boxed{\pi}$ key on a scientific calculator is pressed (on some models, the $\boxed{\text{2nd}}$ key must be pressed first), an approximation of π is displayed. To illustrate how to use this key, consider the following problem. How many times does the tire shown below revolve when a car makes a 25-mile trip?

We first find the circumference of the tire. From the figure, we see that the diameter of the tire is 15 inches. Since the circumference of a circle is the product of π and the length of its diameter, the tire's circumference is $\pi \cdot 15$ inches, or 15π inches. (Normally, we rewrite a product such as $\pi \cdot 15$ so that π is the second factor.)

We then change the 25 miles to inches using two unit conversion factors.

$$\frac{25 \text{ miles}}{1} \cdot \frac{5{,}280 \text{ feet}}{1 \text{ mile}} \cdot \frac{12 \text{ inches}}{1 \text{ foot}} = 25 \cdot 5{,}280 \cdot 12 \text{ inches} \qquad \text{The units of miles and feet can be removed.}$$

The length of the trip is $25 \cdot 5{,}280 \cdot 12$ inches.

Finally, we divide the length of the trip by the circumference of the tire to get

$$\text{The number of revolutions of the tire} = \frac{25 \cdot 5{,}280 \cdot 12}{15\pi}$$

We can use a scientific calculator to make this calculation.

Keystrokes $\boxed{(}\ 25\ \boxed{\times}\ 5280\ \boxed{\times}\ 12\ \boxed{)}\ \boxed{\div}\ \boxed{(}\ 15\ \boxed{\times}\ \boxed{\pi}\ \boxed{)}\ \boxed{=}$

$$\boxed{33613.52398}$$

The tire makes about 33,614 revolutions.

EXAMPLE 2   The circumference of a circle is 50 inches. What is its radius? Give the exact answer and an approximation.

Strategy
We will substitute 50 for C in the formula $C = 2\pi r$ and then solve for r.

Why
The variable r represents the unknown radius of the circle.

Solution

$C = 2\pi r$	This is the formula for the circumference of a circle.
$50 = 2\pi r$	Substitute 50 for C, the circumference of the circle.
$\dfrac{50}{2\pi} = \dfrac{2\pi r}{2\pi}$	To isolate r, undo the multiplication by 2π by dividing both sides by 2π.
$\dfrac{50}{2\pi} = r$	Simplify the right side.
$\dfrac{\overset{1}{2} \cdot 25}{\underset{1}{2\pi}} = r$	To simplify the left side, factor 50 as $2 \cdot 25$ and remove the common factor of 2 in the numerator and denominator.
$\dfrac{25}{\pi} = r$	This is the exact value of r.
$r \approx 7.957747155$	Use a calculator to do the division.

The radius of the circle is approximately 8 inches.

Self Check 2
The circumference of a circle is 25 inches. What is its radius? Give the exact answer and an approximation.

Answer: $\dfrac{25}{2\pi}$ in. ≈ 4 in.

Now Try
Problem 49   ■

EXAMPLE 3   *Architecture.*   A Norman window is constructed by adding a semicircular window to the top of a rectangular window. Find the perimeter of the Norman window shown on the next page.

Strategy
We will find the perimeter of the rectangular part and the circumference of the circular part of the window and add the results.

Why

The window is a combination of a rectangle and a semicircle.

Solution

The perimeter of the rectangular part is

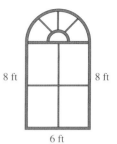

$$P_{\text{rectangular part}} = 8 + 6 + 8 = 22 \quad \text{Add only 3 sides of the rectangle.}$$

The perimeter of the semicircle is one-half of the circumference of a circle that has a 6-foot diameter.

$$P_{\text{semicircle}} = \frac{1}{2}C \qquad \text{This is the formula for the circumference of a semicircle.}$$

$$P_{\text{semicircle}} = \frac{1}{2}\pi D \qquad \text{Since we know the diameter, replace } C \text{ with } \pi D. \text{ We could also have replaced } C \text{ with } 2\pi r.$$

$$= \frac{1}{2}\pi(6) \qquad \text{Substitute 6 for } D, \text{ the diameter.}$$

$$\approx 9.424777961 \qquad \text{Use a calculator to do the multiplication.}$$

The total perimeter is the sum of the two parts.

$$P_{\text{total}} \approx 22 + 9.424777961$$
$$\approx 31.424777961$$

To the nearest hundredth, the perimeter of the window is 31.42 feet.

Now Try

Problem 53

3. Find the area of a circle.

If we divide the circle shown in figure (a) below into an even number of pie-shaped pieces and then rearrange them as shown in figure (b), we have a figure that looks like a parallelogram. The figure has a base b that is one-half the circumference of the circle, and its height h is about the same length as a radius of the circle.

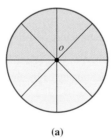   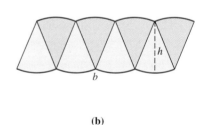

(a)   (b)

If we divide the circle into more and more pie-shaped pieces, the figure will look more and more like a parallelogram, and we can find its area by using the formula for the area of a parallelogram.

$$A = bh$$

$$A = \frac{1}{2}Cr \qquad \text{Substitute } \frac{1}{2} \text{ of the circumference for } b, \text{ the length of the base of the "parallelogram." Substitute } r \text{ for the height of the "parallelogram."}$$

$$= \frac{1}{2}(2\pi r)r \qquad \text{Substitute } 2\pi r \text{ for } C.$$

$$= \pi r^2 \qquad \text{Simplify: } \frac{1}{2} \cdot 2 = 1 \text{ and } r \cdot r = r^2.$$

This result gives the following formula.

Area of a circle	The area of a circle with radius *r* is given by the formula $$A = \pi r^2$$

EXAMPLE 4   Find the area of the circle in the figure on the right. Give the exact answer and an approximation to the nearest tenth.

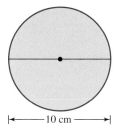

|← 10 cm →|

Strategy
We will find the radius of the circle, substitute that value for *r* in the formula $A = \pi r^2$, and evaluate the right side.

Why
The variable *A* represents the unknown area of the circle.

Solution
Since the length of the diameter is 10 centimeters and the length of a diameter is twice the length of a radius, the length of the radius is 5 centimeters.

$A = \pi r^2$	This is the formula for the area of a circle.
$A = \pi(5)^2$	Substitute 5 for *r*, the radius of the circle. The notation πr^2 means $\pi \cdot r^2$.
$= \pi(25)$	Evaluate the exponential expression.
$= 25\pi$	Write the product so that π is the last factor.

The exact area of the circle is 25π cm². We can use a calculator to approximate the area.

$A \approx 78.53981634$ Use a calculator to do the multiplication: $25 \cdot \pi$.

To the nearest tenth, the area is 78.5 cm².

Self Check 4
Find the area of a circle with a diameter of 12 feet. Give the exact answer and an approximation to the nearest tenth.

Answer: 36π ft² ≈ 113.1 ft²

Now Try
Problem 57

Using Your Calculator: **Painting a helicopter landing pad**

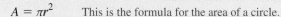

Orange paint is available in gallon containers at $19 each, and each gallon will cover 375 ft². To calculate how much the paint will cost to cover a circular helicopter landing pad 60 feet in diameter, we first calculate the area of the helicopter pad.

$A = \pi r^2$	This is the formula for the area of a circle.
$A = \pi(30)^2$	Substitute one-half of 60 for *r*, the radius of the circular pad.
$= 30^2\pi$	Evaluate the exponential expression.

The area of the pad is exactly $30^2\pi$ ft². Since each gallon of paint will cover 375 ft², we can find the number of gallons of paint needed by dividing $30^2\pi$ by 375.

$$\text{Number of gallons needed} = \frac{30^2\pi}{375}$$

We can use a scientific calculator to make this calculation.

Keystrokes   30 $\boxed{x^2}$ $\boxed{\times}$ $\boxed{\pi}$ $\boxed{=}$ $\boxed{\div}$ 375 $\boxed{=}$ $\boxed{7.539822369}$

Because paint comes only in full gallons, the painter will need to purchase 8 gallons. The cost of the paint will be 8($19), or $152.

EXAMPLE 5   *Crop circles.*   Geometric shapes like that shown on the right have been appearing in the fields of England since the mid-1970s. Since then, *crop circles,* as they are called, have appeared in over 20 countries. If one crop circle was reported to cover an area of 70,000 ft², what was its diameter?

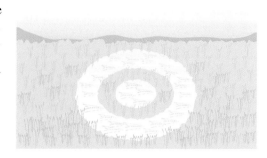

Strategy

We will find the radius of the crop circle by substituting 70,000 for A in the formula $A = \pi r^2$ and then solving for r.

Why

To find the diameter of the crop circle, we simply multiply the radius by 2.

Solution

$A = \pi r^2$ This is the formula for the area of a circle.

$\mathbf{70,000} = \pi r^2$ Substitute 70,000 for A, the area of the crop circle.

$\dfrac{70,000}{\pi} = \dfrac{\pi r^2}{\pi}$ To isolate r^2, undo the multiplication by π by dividing both sides by π.

$\dfrac{70,000}{\pi} = r^2$

To find r, we must find a number that, when squared, is $\dfrac{70,000}{\pi}$. There are two such numbers, one positive and one negative; they are the square roots of $\dfrac{70,000}{\pi}$. Since r is the radius of a circle, r cannot be negative. For this reason, we only find the positive square root of $\dfrac{70,000}{\pi}$ to determine r.

$\sqrt{\dfrac{70,000}{\pi}} = r$ This is the exact value of r. The units are feet.

$r \approx 149.270533$ Use a calculator to find the square root.

To find the diameter of the crop circle, we multiply the radius by 2. We can use a calculator to perform the multiplication.

$D = 2r \approx 2(\mathbf{149.270533}) \approx 298.5$

The diameter of the crop circle is approximately 300 ft.

Now Try

Problem 61

EXAMPLE 6   Find the area of the shaded figure on the right. Round to the nearest hundredth.

Strategy

We will find the area of the entire shaded figure using the following approach:

$A_{\text{total}} = A_{\text{triangle}} + A_{\text{smaller semicircle}} + A_{\text{larger semicircle}}$

Why

The shaded figure is a combination of a triangular region and two semi-circular regions.

Solution

The area of the triangle is

$A_{\text{triangle}} = \frac{1}{2}bh = \frac{1}{2}(\mathbf{6})(\mathbf{8}) = \frac{1}{2}(48) = 24$

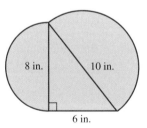

8 in.   10 in.

6 in.

Self Check 6

Find the area of the shaded figure below. Round to the nearest hundredth.

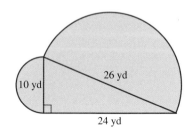

10 yd   26 yd

24 yd

Since the formula for the area of a circle is $A = \pi r^2$, the formula for the area of a semicircle is $A = \frac{1}{2}\pi r^2$. Thus, the area enclosed by the smaller semicircle is

$$A_{\text{smaller semicircle}} = \frac{1}{2}\pi r^2 = \frac{1}{2}\pi(4)^2 = \frac{1}{2}\pi(16) = 8\pi$$

The area enclosed by the larger semicircle is

$$A_{\text{larger semicircle}} = \frac{1}{2}\pi r^2 = \frac{1}{2}\pi(5)^2 = \frac{1}{2}\pi(25) = 12.5\pi$$

The total area is the sum of the three results:

$$A_{\text{total}} = 24 + 8\pi + 12.5\pi \approx 88.4026494 \quad \text{Use a calculator to perform the operations.}$$

To the nearest hundredth, the area of the shaded figure is 88.40 in.^{2}.

Answer: 424.73 yd^2

Now Try
Problem 65

4. Find the length of a circular arc.

A **central angle** of a circle is an angle whose vertex is the center of the circle. In the figure below, the circle has center C. For this circle, $\angle ACB$, with vertex C, is a central angle.

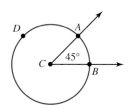

In the figure, the sides of central angle $\angle ACB$ intersect the circle to create minor arc $\overarc{AB}$ and major arc $\overarc{ADB}$. The **degree measure of an arc** is defined to be equal to the degree measure of its corresponding central angle. Since m($\angle ACB$) = 45°, the measure of $\overarc{AB}$ is 45°, and we can write m($\overarc{AB}$) = 45°.

Recall that an angle of one revolution measures 360°. We can use this fact to find the measure of major arc $\overarc{ADB}$ in the figure. In general, the measure of a major arc is simply the difference of 360° and the measure of its associated minor arc.

$$\text{m}(\overarc{ADB}) = 360° - \text{m}(\overarc{AB}) = 360° - 45° = 315°$$

We can use the following formula to find the length of an arc of a circle.

Length of an arc

If an arc has measure q (in degrees) and radius r, its length L is given by

$$L = \frac{q}{360°} \cdot 2\pi r \quad L \text{ and } r \text{ have the same units.}$$

Note that the length of an arc of a circle is a fraction of the circumference of the circle ($2\pi r$). The fraction is the ratio of the measure of the arc to one complete revolution, 360°.

EXAMPLE 7   Find the length of minor arc $\overarc{AB}$ shown on the right. C is the center of the circle. Give the exact answer and an approximation to the nearest tenth.

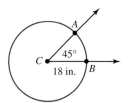

Strategy
$\overarc{AB}$ corresponds to central angle $\angle ACB$. Since m($\angle ACB$) = 45°, m($\overarc{AB}$) is also 45°. We will substitute 45° for q and 18 for r in the formula $L = \frac{q}{360°} \cdot 2\pi r$ and evaluate the right side.

Why
The variable L represents the unknown arc length.

Self Check 7
Find the length of minor arc $\overarc{MN}$ shown below. C is the center of the circle. Give the exact answer and an approximation to the nearest tenth.

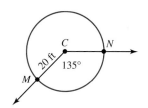

Solution

$$L = \frac{q}{360°} \cdot 2\pi r$$ This is the formula for the arc length of a circle.

$$L = \frac{45°}{360°} \cdot 2\pi(18)$$ Substitute 45° for q, the measure of the arc. Substitute 18 for r, the radius.

$$= \frac{\overset{1}{45°}}{8 \cdot \underset{1}{45°}} \cdot 2\pi(18)$$ To simplify the fraction, factor 360° as 8 · 45° and remove the common factor 45°.

$$= \frac{1}{8} \cdot 36\pi$$ Multiply $2\pi(18)$ and write the product so that π is the last factor: 36π. Note that the length of the arc is $\frac{1}{8}$ of the circumference of the circle, 36π.

$$= \frac{36\pi}{8}$$ Multiply: $\frac{1}{8} \cdot 36\pi = \frac{1}{8} \cdot \frac{36\pi}{1}$.

$$= \frac{\overset{1}{4} \cdot 9\pi}{2 \cdot \underset{1}{4}}$$ Factor 36π as 4 · 9π and 8 as 2 · 4. To simplify, remove the common factor of 4 in the numerator and denominator.

$$= \frac{9\pi}{2}$$

The length of $\overset{\frown}{AB}$ is exactly $\frac{9\pi}{2}$ inches. To the nearest tenth, $\frac{9\pi}{2}$ inches ≈ 14.1 inches.

Answer: 15π ft ≈ 47.1 ft

Now Try
Problem 69   ■

5. Find the area of a sector.

The shaded region in the figure below is called a **sector.** The sector in the figure has a radius of 20 meters, and the measure of its associated arc is 60°.

We can find the area of the sector using the following formula.

Area of a sector   If a sector has radius r, and its associated arc has measure q, its area is given by

$$A = \frac{q}{360°} \cdot \pi r^2$$

Note that the area of a sector is a fraction of the area of the circle (πr^2). The fraction is the ratio of the associated arc measure to one complete revolution, 360°.

EXAMPLE 8   Find the area of sector shown on the right. C is the center of the circle. Give the exact answer and an approximation to the nearest tenth.

Strategy
$\overset{\frown}{AB}$ corresponds to central angle $\angle ACB$. Since m($\angle ACB$) = 60°, m($\overset{\frown}{AB}$) is also 60°. We will substitute 60° for q and 20 for r in the formula $A = \frac{q}{360°} \cdot \pi r^2$ and evaluate the right side.

Why
The variable A represents the unknown area of the sector.

Self Check 8
Find the area of the sector shown below. Give the exact answer and an approximation to the nearest tenth.

Solution

$A = \dfrac{q}{360°} \cdot \pi r^2$ This is the formula for the area of a sector.

$A = \dfrac{\mathbf{60°}}{360°} \cdot \pi (\mathbf{20})^2$ Substitute 60° for *q*, the measure of the arc. Substitute 20 for *r*, the radius.

$A = \dfrac{60°}{360°} \cdot \pi (400)$ Evaluate the exponential expression: $(20)^2 = 400$.

$A = \dfrac{\overset{1}{\cancel{60°}}}{6 \cdot \cancel{60°}} \cdot 400\pi$ To simplify the fraction, factor 360° as $6 \cdot 60°$ and remove the common factor 60°. Write $\pi(400)$ so that π is the last factor: 400π.

$= \dfrac{1}{6} \cdot 400\pi$ Note that the area of the sector is $\frac{1}{6}$ of the area of the circle, 400π.

$= \dfrac{400\pi}{6}$ Multiply: $\frac{1}{6} \cdot 400\pi = \frac{1}{6} \cdot \frac{400\pi}{1}$.

$= \dfrac{\overset{1}{\cancel{2}} \cdot 200\pi}{\underset{1}{\cancel{2}} \cdot 3}$ To simplify the fraction, factor 400π as $2 \cdot 200\pi$ and 6 as $2 \cdot 3$. Remove the common factor 2 in the numerator and denominator.

$= \dfrac{200\pi}{3}$

The area of the sector is exactly $\frac{200\pi}{3}$ square meters. To the nearest tenth, $\frac{200\pi}{3}$ m² ≈ 209.4 m².

Answer: $\frac{32\pi}{9}$ cm² ≈ 11.2 cm²

Now Try
Problem 73

STUDY SET   Section 7

VOCABULARY   *Fill in the blanks.*

1. A segment drawn from the center of a circle to a point on the circle is called a _____.

2. A segment joining two points on a circle is called a _____.

3. A _____ is a chord that passes through the center of a circle.

4. An arc that is one-half of a complete circle is a _____.

5. The distance around a circle is called its _____.

6. The surface enclosed by a circle is called its _____.

7. A diameter of a circle is _____ as long as a radius.

8. Suppose the exact circumference of a circle is 3π feet. When we write $C \approx 9.42$ feet, we are giving an _____ of the circumference.

9. An arc that is shorter than a semicircle is called a _____ arc. An arc that is longer than a semicircle is called a _____ arc.

10. A _____ angle of a circle is an angle whose vertex is the center of the circle.

11. The degree measure of an _____ is defined to be the degree measure of its corresponding central angle.

12. The shaded region shown below is called a _____.

CONCEPTS   *Refer to the figure below, where point O is the center of the circle.*

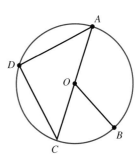

13. Name each radius.

14. Name a diameter.

15. Name each chord.

16. Name each minor arc.

17. Name each semicircle.

18. Name major arc $\overset{\frown}{ABD}$ in another way.

19. The sides of what central angle intersect the circle to create minor arc $\overset{\frown}{BC}$?

20. The sides of what central angle intersect the circle to create major arc $\overset{\frown}{ADB}$?

21. a. If you know the radius of a circle, how can you find its diameter?

  b. If you know the diameter of a circle, how can you find its radius?

22. One complete revolution is how many degrees?

23. What are the two formulas that can be used to find the circumference of a circle?

24. What is the formula for the area of a circle?

25. If C is the circumference of a circle and D is its diameter, then $\frac{C}{D} = $ ___ .

26. If D is the diameter of a circle and r is its radius, then $D = $ ___ r.

27. When evaluating $\pi(6)^2$, what operation should be performed first?

28. Round $\pi = 3.141592653589\ldots$ to the nearest hundredth.

Refer to the figure below, where point X is the center of the circle.

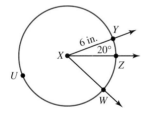

29. What is the diameter of the circle?

30. What is m($\overset{\frown}{YZ}$)? What is m($\overset{\frown}{YUZ}$)?

31. Name $\overset{\frown}{YUZ}$ in another way.

32. What is m($\overset{\frown}{YWZ}$)?

33. If m($\overset{\frown}{ZW}$) = 42°, what is m($\angle ZYW$)?

34. If m($\overset{\frown}{ZW}$) = 42°, what is m($\overset{\frown}{ZUW}$)?

35. What is the central angle associated with minor arc $\overset{\frown}{YW}$?

36. What is the radius of the sector associated with central angle $\angle ZXW$?

37. a. On the given circle, draw central angle $\angle ABC$ and label it completely.

b. What is the formula that gives the length of $\overset{\frown}{AC}$?

  c. What part of the formula represents the entire circumference of the circle?

  d. What fraction of the entire circumference does this formula find?

38. a. On the given circle, draw central angle $\angle ABC$; then shade the sector that is associated with it. Label the figure completely.

  b. What is the formula that gives the area of the sector?

  c. What part of the formula represents the entire area of the circle?

  d. What fraction of the entire area does this formula find?

39. Write each expression in better form. Leave π in your answer.

  a. $\pi(8)$ **b.** $2\pi(7)$ **c.** $\pi \cdot \frac{25}{3}$

40. Simplify each fraction. Leave π in your answer.

  a. $\frac{90°}{360°}$ **b.** $\frac{4\pi}{8}$ **c.** $\frac{27\pi}{30}$

NOTATION *Fill in the blanks.*

41. The symbol $\overset{\frown}{AB}$ is read as "___ ___ ."

42. To the nearest hundredth, the value of π is _____ .

43. a. In the expression $2\pi r$, what operations are indicated?

  b. In the expression πr^2, what operations are indicated?

44. a. What does m($\overset{\frown}{AB}$) mean?

  b. What does m($\overset{\frown}{DEF}$) mean?

GUIDED PRACTICE *The answers may vary slightly, depending on which approximation of π is used.*

Find the circumference of the circle shown below. Give the exact answer and an approximation to the nearest tenth. **See Example 1.**

45.
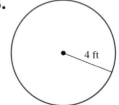

46.
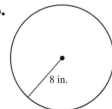

47.
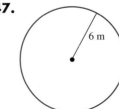

48.

Find the radius of each circle given its circumference. Give the exact answer and an approximation to the nearest hundredth. **See Example 2.**

49. The circumference is 24 in.

50. The circumference is 34 yd.

51. The circumference is 18 cm.

52. The circumference is 100 cm.

Find the perimeter of each figure. Assume each arc is a semicircle. Round to the nearest hundredth. **See Example 3.**

53.
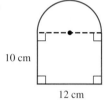
10 cm
12 cm

54.
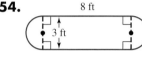
8 ft
3 ft

55.
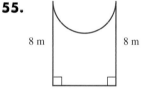
8 m   8 m
6 m

56.
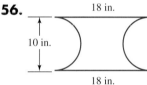
18 in.
10 in.
18 in.

Find the area of each circle given the following information. Give the exact answer and an approximation to the nearest tenth. **See Example 4.**

57.

6 in.

58.
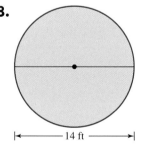
14 ft

59. Find the area of a circle with diameter 18 inches.

60. Find the area of a circle with diameter 20 meters.

Find the diameter of each circle given its area. Round to the nearest tenth. **See Example 5.**

61. The area is 280 ft^2.

62. The area is 500 m^2.

63. The area is 9,900 in.^{2}.

64. The area is 16,800 mi^2.

Find the total area of each figure. Assume each arc is a semicircle. Round to the nearest tenth. **See Example 6.**

65.
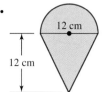
12 cm
12 cm

66.
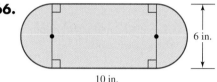
6 in.
10 in.

67.
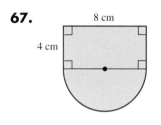
8 cm
4 cm

68.
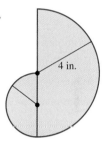
4 in.

Find the length of minor arc $\overarc{AB}$. Give the exact answer and an approximation to the nearest tenth. **See Example 7.**

69.
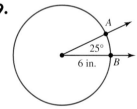
A
25°
6 in.   B

70.
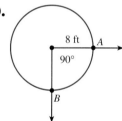
8 ft   A
90°
B

71.
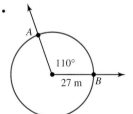
A
110°
27 m   B

72.
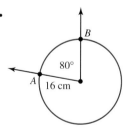
B
80°
A   16 cm

Find the area of the shaded sector. Give the exact answer and an approximation to the nearest tenth. **See Example 8.**

73.
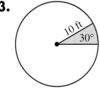
10 ft
30°

74.

45°
9 in.

75.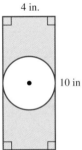
150°
30 cm

76.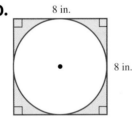
50°
18 mi

TRY IT YOURSELF

The answers may vary slightly, depending on which approximation of π is used.

77. The area of a circle is 4.4 m². Round each of the following answers to the nearest tenth.

   a. What is its radius?

   b. What is its diameter?

   c. What is its circumference?

78. The area of a circle is 150 cm². Round each of the following answers to the nearest tenth.

   a. What is its radius?

   b. What is its diameter?

   c. What is its circumference?

Find the area of each shaded region. Round to the nearest tenth.

79.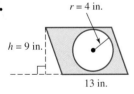
4 in.
10 in

80.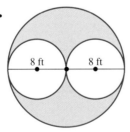
8 in.
8 in.

81.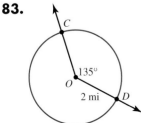
r = 4 in.
h = 9 in.
13 in.

82.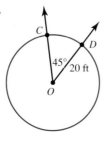
8 ft   8 ft

Find the length of minor arc $\overparen{CD}$. Give the exact answer and an approximation to the nearest tenth.

83.
C
135°
O
2 mi   D

84.
C
D
45°
20 ft
O

85. Find the circumference of the circle shown below. Give the exact answer and an approximation to the nearest hundredth.

50 yd

86. Find the circumference of the semicircle shown below. Give the exact answer and an approximation to the nearest hundredth.

25 cm

87. Find the radius of a circle that has an area of 49π ft².

88. Find the radius of a circle that has an area of 64π cm².

89. Find the circumference of the circle shown below if the square has sides of length 6 inches. Give the exact answer and an approximation to the nearest tenth.

90. Find the circumference of the semicircle shown below if the length of the rectangle is 8 feet. Give the exact answer and an approximation to the nearest hundredth.

Find the area of the shaded sector. Give the exact answer and an approximation to the nearest tenth.

91.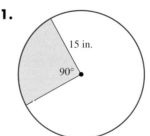
15 in.
90°

92.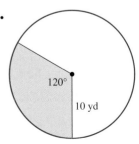
120°
10 yd

93. Find the circumference of a circle that has a diameter of 120 inches. Give the exact answer and an approximation to the nearest tenth.

94. Find the circumference of a circle that has a diameter of 46 feet. Give the exact answer and an approximation to the nearest tenth.

95. Find the diameter of a circle that has a circumference of 113 meters. Give the exact answer and an approximation to the nearest hundredth.

96. Find the diameter of a circle that has a circumference of 157 meters. Give the exact answer and an approximation to the nearest hundredth.

97. Find the area of the circle shown below if the square has sides of length 9 millimeters. Give the exact answer and an approximation to the nearest tenth.

98. Find the area of the shaded semicircular region shown below. Give the exact answer and an approximation to the nearest tenth.

| ← 6.5 mi → |

99. Find the radius of a circle that has a circumference of 16π inches.

100. Find the radius of a circle that has a circumference of 30π meters.

101. Find the diameter of a circle that has a circumference of 5π centimeters.

102. Find the diameter of a circle that has a circumference of 9π yards.

103. Find the area of a circle with radius 15 feet. Give the exact answer and an approximation to the nearest hundredth.

104. Find the area of a circle with radius 3.2 inches. Give the exact answer and an approximation to the nearest hundredth.

105. Find the diameter of a circle that has an area of $\frac{25\pi}{16}$ yd².

106. Find the diameter of a circle that has an area of $\frac{36\pi}{25}$ mi².

APPLICATIONS

107. Suppose the two "legs" of the compass shown below are adjusted so that the distance between the pointed ends is 1 inch. Then a circle is drawn.

   a. What will the radius of the circle be?

   b. What will the diameter of the circle be?

   c. What will the circumference of the circle be? Give an exact answer and an approximation to the nearest hundredth.

   d. What will the area of the circle be? Give an exact answer and an approximation to the nearest hundredth.

108. Suppose we find the distance around a can and the distance across the can using a measuring tape, as shown below. Then we make a comparison, in the form of a ratio:

$$\frac{\text{The distance around the can}}{\text{The distance across the top of the can}}$$

After we do the indicated division, the result will be close to what number?

When appropriate, give the exact answer and an approximation to the nearest hundredth. Answers may vary slightly, depending on which approximation of π is used.

109. LAKES Round Lake has a circular shoreline that is 2 miles in diameter. Find the area of the lake.

110. HELICOPTERS Refer to the figure below. How far does a point on the tip of a rotor blade travel when it makes one complete revolution?

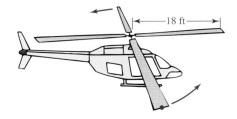

111. GIANT SEQUOIA   The largest sequoia tree is the General Sherman Tree in Sequoia National Park in California. In fact, it is considered to be the largest living thing in the world. According to the *Guinness Book of World Records,* it has a circumference of 102.6 feet, measured $4\frac{1}{2}$ feet above the ground. What is the diameter of the tree at that height?

112. TRAMPOLINE   See the figure below. The distance from the center of the trampoline to the edge of its steel frame is 7 feet. The protective padding covering the springs is 18 inches wide. Find the area of the circular jumping surface of the trampoline, in square feet.

Protective
pad

113. JOGGING   Joan wants to jog 10 miles on a circular track $\frac{1}{4}$ mile in diameter. How many times must she circle the track? Round to the nearest lap.

114. CARPETING   A state capitol building has a circular floor 100 feet in diameter. The legislature wishes to have the floor carpeted. The lowest bid is $83 per square yard, including installation. How much must the legislature spend for the carpeting project? Round to the nearest dollar.

115. BANDING THE EARTH   A steel band is drawn tightly about the Earth's equator. The band is then loosened by increasing its length by 10 feet, and the resulting slack is distributed evenly along the band's entire length. How far above the Earth's surface is the band? (*Hint:* You don't need to know the Earth's circumference.)

116. CONCENTRIC CIRCLES   Two coplanar circles are called **concentric circles** if they have the same center. Find the area of the band between two concentric circles if their diameters are 10 centimeters and 6 centimeters.

117. ARCHERY   See the figure below. Find the area of the entire target and the area of the bull's eye. What percent of the area of the target is the bull's eye?

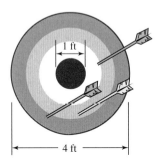

1 ft

4 ft

118. LANDSCAPE DESIGN   See the figure below. How many square feet of lawn does not get watered by the four sprinklers at the center of each circle?

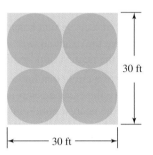

30 ft

30 ft

119. AUTOMOTIVE REPAIR   The figure below shows how a fan belt turns pulleys connected to a car's alternator and water pump.

   a. How many inches of the fan belt touch the alternator pulley?

   b. How many inches of the fan belt touch the water pump pulley?

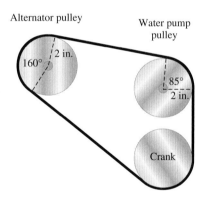

Alternator pulley

Water pump pulley

160°   2 in.

85°
2 in.

Crank

120. CLOCKS   The figure below shows the minute hand of a clock moving from 12 to 3.

   a. What angle does the minute hand sweep out?

   b. If the minute hand is 5 inches long, how much area does it sweep out?

WRITING

121. Explain what is meant by the circumference of a circle.

122. Explain what is meant by the area of a circle.

123. Explain the meaning of π.

124. Distinguish between a major arc and a minor arc.

125. Explain what it means for a car to have a small turning radius.

126. The word *circumference* means the distance around a circle. In your own words, explain what is meant by each of the following sentences.

 a. A boat owner's dream was to *circumnavigate* the globe.

 b. The teenager's parents felt that he was always trying to *circumvent* the rules.

 c. The class was shown a picture of a circle *circumscribed* about an equilateral triangle.

Sections 1-7 Cumulative Review Problems

1. What are the three undefined words in geometry?

2. Draw each geometric figure and label it.

    a. $\overleftrightarrow{AS}$ **b.** $\overline{GH}$

    c. $\overrightarrow{MN}$ **d.** $\angle BCA$

3. Are $\overrightarrow{AB}$ and $\overrightarrow{BA}$ the same ray?

4. Give four ways to name the angle shown below.

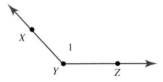

5. Measure each angle with a protractor. Then tell whether it is an acute, right, obtuse, or straight angle.

    a.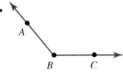

    b.

    c.

    d.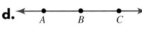

6. Fill in the blanks.

    a. If $\angle ABC \cong \angle DEF$, then the angles have the same _____.

    b. Two congruent segments have the same _____.

    c. Two different points determine one _____.

    d. Two angles are called _____ if the sum of their measures is 90°.

7. Refer to the figure below. What is the midpoint of $\overline{BE}$?

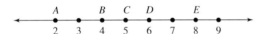

8. Refer to the figure below and tell whether each statement is true or false.

    a. $\angle AGF$ and $\angle BGC$ are vertical angles.

    b. $\angle EGF$ and $\angle DGE$ are adjacent angles.

    c. $m(\angle AGB) = m(\angle EGD)$.

    d. $\angle CGD$ and $\angle DGF$ are supplementary angles.

    e. $\angle EGD$ and $\angle AGB$ are complementary angles.

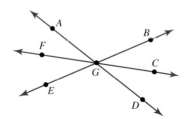

9. Find x. Then find $m(\angle ABD)$ and $m(\angle CBE)$.

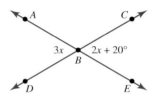

10. The measure of an angle is 20° more than three times its supplement's measure. What is the measure of the angle?

11. Refer to the figure below. Fill in the blanks.

    a. l_1 intersects two coplanar lines. It is called a _____.

    b. $\angle 4$ and _____ are alternate interior angles.

    c. $\angle 3$ and _____ are corresponding angles.

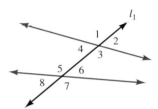

12. In the figure below, $l_1 \parallel l_2$ and $m(\angle 2) = 25°$. Find the measures of the other numbered angles.

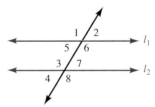

13. In the figure below, $l_1 \parallel l_2$. Find x. Then determine the measure of each angle that is labeled in the figure below.

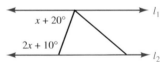

14. Are l_1 and l_2 parallel? If so, explain why.

15. For each polygon, give the number of sides it has, tell its name, and then give the number of vertices it has.

a.

b.

c.

d.

16. Classify each triangle as an equilateral triangle, an isosceles triangle, or a scalene triangle.

a.

b.
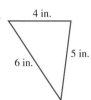
4 in.
5 in.
6 in.

c.
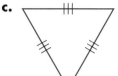

d.

56°
56°

17. Find *x*.

x
20°

18. The measure of each base angle of an isosceles triangle is 30° more than the measure of the vertex angle. Find the measure of each angle.

19. LOGOS   What three quadrilaterals are used in the design of the Chevrolet ad shown below?

Genuine Chevrolet

20. Draw each figure.

a. Rhombus *ABCD*   **b.** Trapezoid *QRST*

21. Refer to rectangle *EFGH* shown below.

a. Find m($\overline{HG}$).   **b.** Find m($\overline{FH}$).

c. Find m($\angle FGH$).   **d.** Find m($\overline{EH}$).

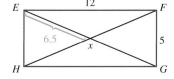
E   12   *F*
6.5
x
5
H   *G*

22. Refer to isosceles trapezoid *QRST* shown below.

a. Find m($\overline{RS}$).   **b.** Find *x*.

c. Find *y*.   **d.** Find *z*.

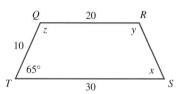
Q   20   *R*
z   *y*
10
65°   *x*
T   30   *S*

23. Find the sum of the measures of the angles of a decagon.

24. a. Find the number of sides of a regular polygon if one of its angles has a measure of 120°.

b. What is the measure of an exterior angle of the polygon in part a?

25. Find the perimeter of the figure shown below.

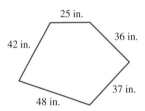
25 in.
42 in.   36 in.
37 in.
48 in.

26. The perimeter of an equilateral triangle is 45.6 m. Find the length of each side.

27. The perimeter of a rectangle is 34 ft. The length is 3 feet less than four times the width. Find the length and width.

28. Find the area of the shaded part of the figure shown below.

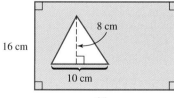
8 cm
16 cm
10 cm
25 cm

29. The dimensions of the parallelogram shown below are represented by algebraic expressions. In each case, the units are feet. Write an algebraic expression that represents the area of the parallelogram.

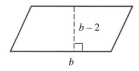
b − 2
b

30. Draw an altitude to the base of the triangle shown below.

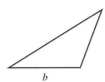

b

31. DECORATING   A patio has the shape of a trapezoid, as shown below. If indoor/outdoor carpeting sells for $18 a square yard installed, how much will it cost to carpet the patio?

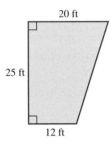

20 ft

25 ft

12 ft

32. How many square inches are in one square foot?

33. A student said the area of the square shown below was 25^2 ft. Is this correct? Explain why or why not.

5 ft

5 ft

34. Find the area of the rectangle shown below.

10 ft

1 in.

35. Refer to the figure below, where O is the center of the circle.

 a. Name each chord.

 b. Name each diameter.

 c. Name each radius.

 d. The sides of central angle $\angle DOB$ intersect the circle to form what minor and what major arc?

 e. What type of figure is $\overset{\frown}{ACB}$?

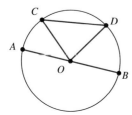

36. Fill in the blank: If C is the circumference of a circle and D is the length of its diameter, then $\frac{C}{D} = $ _____ .

In Problems 37–46, when appropriate, give the exact answer and an approximation to the nearest tenth.

37. Find the circumference of a circle with a diameter of 21 centimeters.

38. Find the perimeter of the figure shown below. Assume that the arcs are semicircles.

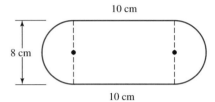

10 cm

8 cm

10 cm

39. TOYS   The circumference of a Hula Hoop is 106.8 inches. What is its radius?

40. HISTORY   Stonehenge is a prehistoric monument in England, believed to have been built by the Druids. The site, 30 meters in diameter, consists of a circular arrangement of stones, as shown in below. What area does the monument cover?

41. Find the area of the shaded part of the figure shown below, if the square has sides of length 4.5 miles.

42. DANCE FLOORS   Brown paint is available in gallon containers at $28 each, and each gallon will cover 350 ft². How much will the paint cost to cover a circular dance floor that is 40 feet in diameter?

43. HOT TUB   The surface area of the water in a circular hot tub is 38.5 ft².

 a. What is its radius? Give the exact answer and an approximation to the nearest tenth.

 b. What is its diameter? Round to the nearest tenth.

 c. What is its circumference? Round to the nearest tenth.

44. Refer to the figure below, where *A* is the center of the circle.

   a. What is m($\overarc{XY}$)?

   b. What is m($\overarc{YZX}$)?

   c. What is m($\overarc{YZ}$)?

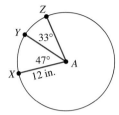

45. Find the length of $\overarc{YZ}$ shown in the figure for Problem 44. Give the exact answer and an approximation to the nearest tenth.

46. SNOWMAKING A snowmaking machine sends out a spray that covers an area shaped like a sector of a circle, as shown below. Find the area of the sector. Give the exact answer and an approximation to the nearest tenth.

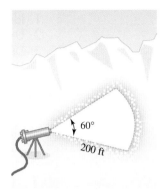

8 *Congruent Triangles and Similar Triangles*

Objectives

1. Identify corresponding parts of congruent triangles.
2. Use congruence properties to prove that two triangles are congruent.
3. Determine whether two triangles are similar.
4. Use similar triangles to find unknown lengths in application problems.

In our everyday lives, we see many types of triangles. Triangular-shaped kites, sails, roofs, tortilla chips, and ramps are just a few examples. In this section, we will discuss how to formally compare the size and shape of two given triangles. From this comparison, we can make deductions about their respective side lengths and angle measures.

1. Identify corresponding parts of congruent triangles.

Simply put, two geometric figures are congruent if they have the same shape and size. For example, if $\triangle ABC$ and $\triangle DEF$ in the figure below are congruent, we can write

$\triangle ABC \cong \triangle DEF$ Read as "Triangle *ABC* is congruent to triangle *DEF*."

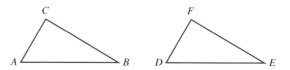

One way to determine whether two triangles are congruent is to see if one triangle can be moved onto the other triangle in such a way that it fits exactly. When we write $\triangle ABC \cong \triangle DEF$, we are showing how the vertices of one triangle are matched to the vertices of the other triangle to obtain a "perfect fit." We call this matching of points a **correspondence.**

$\triangle ABC \cong \triangle DEF$

$A \leftrightarrow D$ Read as "Point *A* corresponds to point *D*."

$B \leftrightarrow E$ Read as "Point *B* corresponds to point *E*."

$C \leftrightarrow F$ Read as "Point *C* corresponds to point *F*."

When we establish a correspondence between the vertices of two congruent triangles, we also establish a correspondence between the angles and the sides of the triangles. Corresponding angles and corresponding sides of congruent triangles are called **corresponding parts.** *Corresponding parts of congruent triangles are always congruent.* That is, corresponding parts of congruent triangles always have the same measure. For the congruent triangles in the figure above, we have

$$\text{m}(\angle A) = \text{m}(\angle D) \qquad \text{m}(\angle B) = \text{m}(\angle E) \qquad \text{m}(\angle C) = \text{m}(\angle F)$$
$$\text{m}(\overline{BC}) = \text{m}(\overline{EF}) \qquad \text{m}(\overline{AC}) = \text{m}(\overline{DF}) \qquad \text{m}(\overline{AB}) = \text{m}(\overline{DE})$$

Congruent triangles | Two triangles are congruent if and only if their vertices can be matched so that the corresponding sides and the corresponding angles are congruent.

EXAMPLE 1   Refer to the figure below, where $\triangle XYZ \cong \triangle PQR$.

a. Name the six congruent corresponding parts of the triangles.

b. Find $m(\angle P)$.

c. Find $m(\overline{XZ})$.

Strategy

We will establish the correspondence between the vertices of $\triangle XYZ$ and the vertices of $\triangle PQR$.

Why

This will, in turn, establish a correspondence between the congruent corresponding angles and sides of the triangles.

Solution

a. The correspondence between vertices is

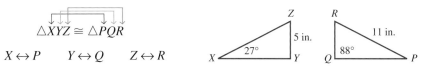

$$\triangle XYZ \cong \triangle PQR$$

$$X \leftrightarrow P \qquad Y \leftrightarrow Q \qquad Z \leftrightarrow R$$

Corresponding parts of congruent triangles are congruent. Therefore, the congruent corresponding angles are

$$\angle X \cong \angle P \qquad \angle Y \cong \angle Q \qquad \angle Z \cong \angle R$$

The congruent corresponding sides are

$$\overline{YZ} \cong \overline{QR} \qquad \overline{XZ} \cong \overline{PR} \qquad \overline{XY} \cong \overline{PQ}$$

b. From the figure, we see that $m(\angle X) = 27°$. Since $\angle X \cong \angle P$, it follows that $m(\angle P) = 27°$.

c. From the figure, we see that $m(\overline{PR}) = 11$ inches. Since $\overline{XZ} \cong \overline{PR}$, it follows that $m(\overline{XZ}) = 11$ inches.

Now Try

Problem 33   ■

2. Use congruence properties to prove that two triangles are congruent.

Sometimes it is possible to conclude that two triangles are congruent without having to show that three pairs of corresponding angles are congruent and three pairs of corresponding sides are congruent. To do so, we apply one of the following properties.

SSS property

> If three sides of one triangle are congruent to three sides of a second triangle, the triangles are congruent.

We can show that the triangles shown below are congruent by the SSS property:

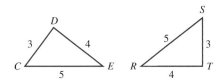

$\overline{CD} \cong \overline{ST}$ Since m($\overline{CD}$) = 3 and m($\overline{ST}$) = 3, the segments are congruent.

$\overline{DE} \cong \overline{TR}$ Since m($\overline{DE}$) = 4 and m($\overline{TR}$) = 4, the segments are congruent.

$\overline{EC} \cong \overline{RS}$ Since m($\overline{EC}$) = 5 and m($\overline{RS}$) = 5, the segments are congruent.

Therefore, $\triangle CDE \cong \triangle STR$.

SAS property

> If two sides and the angle between them in one triangle are congruent, respectively, to two sides and the angle between them in a second triangle, the triangles are congruent.

We can show that the triangles shown below are congruent by the SAS property:

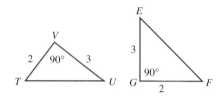

$\overline{TV} \cong \overline{FG}$ Since m($\overline{TV}$) = 2 and m($\overline{FG}$) = 2, the segments are congruent.

$\angle V \cong \angle G$ Since m($\angle V$) = 90° and m($\angle G$) = 90°, the angles are congruent.

$\overline{UV} \cong \overline{EG}$ Since m($\overline{UV}$) = 3 and m($\overline{EG}$) = 3, the segments are congruent.

Therefore, $\triangle TVU \cong \triangle FGE$.

ASA property

> If two angles and the side between them in one triangle are congruent, respectively, to two angles and the side between them in a second triangle, the triangles are congruent.

We can show that the triangles shown below are congruent by the ASA property:

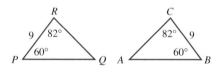

$\angle P \cong \angle B$ Since m($\angle P$) = 60° and m($\angle B$) = 60°, the angles are congruent.

$\overline{PR} \cong \overline{BC}$ Since m($\overline{PR}$) = 9 and m($\overline{BC}$) = 9, the segments are congruent.

$\angle R \cong \angle C$ Since m($\angle R$) = 82° and m($\angle C$) = 82°, the angles are congruent.

Therefore, $\triangle PQR \cong \triangle BAC$.

CAUTION   There is no SSA property. To illustrate this, consider the triangles shown below. Two sides and an angle of $\triangle ABC$ are congruent to two sides and an angle of $\triangle DEF$. But the congruent angle is not between the congruent sides.

We refer to this situation as SSA. Obviously, the triangles are not congruent because they are not the same shape and size.

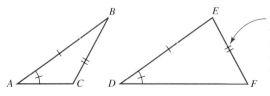

The tick marks indicate congruent parts. That is, the sides with one tick mark are the same length, the sides with two tick marks are the same length, and the angles with one tick mark have the same measure.

EXAMPLE 2 Explain why the triangles in the figure on the right are congruent.

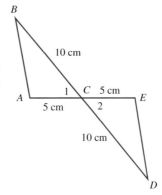

Strategy
We will show that two sides and the angle between them in one triangle are congruent, respectively, to two sides and the angle between them in a second triangle.

Why
Then we know that the two triangles are congruent by the SAS property.

Solution
Since vertical angles are congruent,

$$\angle 1 \cong \angle 2$$

From the figure, we see that

$$\overline{AC} \cong \overline{EC} \quad \text{and} \quad \overline{BC} \cong \overline{DC}$$

Since two sides and the angle between them in one triangle are congruent, respectively, to two sides and the angle between them in a second triangle, $\triangle ABC \cong \triangle EDC$ by the SAS property.

Self Check 2
Are the triangles in the figure below congruent? Explain why or why not.

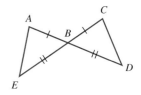

Answer: yes, by the SAS property

Now Try
Problem 35

EXAMPLE 3 Are $\triangle RST$ and $\triangle RUT$ in the figure on the right congruent?

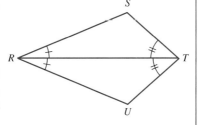

Strategy
We will show that two angles and the side between them in one triangle are congruent, respectively, to two angles and the side between them in a second triangle.

Why
Then we know that the two triangles are congruent by the ASA property.

Solution
From the markings on the figure, we know that two pairs of angles are congruent.

$\angle SRT \cong \angle URT$ These angles are marked with 1 tick mark, which indicates that they have the same measure.

$\angle STR \cong \angle UTR$ These angles are marked with 2 tick marks, which indicates that they have the same measure.

From the figure, we see that the triangles have side $\overline{RT}$ in common. Furthermore, $\overline{RT}$ is between each pair of congruent angles listed above. Since every segment is congruent to itself, we also have

$$\overline{RT} \cong \overline{RT}$$

Knowing that two angles and the side between them in $\triangle RST$ are congruent, respectively, to two angles and the side between them in $\triangle RUT$, we can conclude that $\triangle RST \cong \triangle RUT$ by the ASA property.

Self Check 3
Are the triangles in the following figure congruent? Explain why or why not.

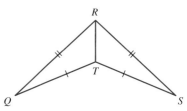

Answer: yes, by the SSS property

Now Try
Problem 37

---

3. Determine whether two triangles are similar.

We have seen that congruent triangles have the same shape and size. Similar triangles have the same shape, but not necessarily the same size. That is, one triangle is an exact scale model of the other triangle. If the triangles in the figure below are similar, we can write $\triangle ABC \sim \triangle DEF$ (read the symbol $\sim$ as "is similar to").

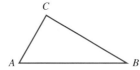
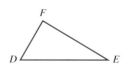

 CAUTION   Note that congruent triangles are always similar, but similar triangles are not always congruent.

The formal definition of similar triangles requires that we establish a correspondence between the vertices of the triangles. The definition also involves the word *proportional.*

Recall that a **proportion** is a mathematical statement that two ratios (fractions) are equal. An example of a proportion is

$$\frac{1}{2} = \frac{4}{8}$$

In this case, we say that $\frac{1}{2}$ and $\frac{4}{8}$ are *proportional.*

Similar triangles

> Two triangles are similar if and only if their vertices can be matched so that corresponding angles are congruent and the lengths of corresponding sides are proportional.

EXAMPLE 4   Refer to the figure on the right. If $\triangle PQR \sim \triangle CDE$, name the congruent angles and the sides that are proportional.

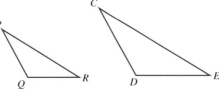

Strategy
We will establish the correspondence between the vertices of $\triangle PQR$ and the vertices of $\triangle CDE$.

Why
This will, in turn, establish a correspondence between the congruent corresponding angles and proportional sides of the triangles.

Solution
When we write $\triangle PQR \sim \triangle CDE$, a correspondence between the vertices of the triangles is established.

$$\triangle PQR \sim \triangle CDE$$

Since the triangles are similar, corresponding angles are congruent:

$$\angle P \cong \angle C \qquad \angle Q \cong \angle D \qquad \angle R \cong \angle E$$

The lengths of the corresponding sides are proportional. (To simplify the notation, we will now let $PQ = m(\overline{PQ})$, $CD = m(\overline{CD})$, $QR = m(\overline{QR})$, and so on.)

$$\frac{PQ}{CD} = \frac{QR}{DE} \qquad \frac{QR}{DE} = \frac{PR}{CE} \qquad \frac{PQ}{CD} = \frac{PR}{CE}$$

Written in a more compact way,

$$\frac{PQ}{CD} = \frac{QR}{DE} = \frac{PR}{CE}$$

Self Check 4

If $\triangle GEF \sim \triangle IJH$, name the congruent angles and the sides that are proportional.

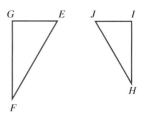

Answer: $\angle G \cong \angle I$, $\angle E \cong \angle J$, $\angle F \cong \angle H$; $\dfrac{EG}{JI} = \dfrac{GF}{IH}$, $\dfrac{GF}{IH} = \dfrac{FE}{HJ}$, $\dfrac{EG}{JI} = \dfrac{FE}{HJ}$

Now Try
Problem 39

Property of similar triangles

> If two triangles are similar, all pairs of corresponding sides are in proportion.

It is possible to conclude that two triangles are similar without having to show that all three pairs of corresponding angles are congruent and that the lengths of all three pairs of corresponding sides are proportional.

AAA similarity theorem

> If the angles of one triangle are congruent to corresponding angles of another triangle, the triangles are similar.

EXAMPLE 5   In the figure on the right, $\overline{PR} \parallel \overline{MN}$. Are $\triangle PQR$ and $\triangle NQM$ similar triangles?

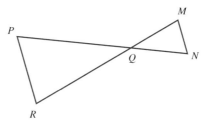

Strategy
We will show that the angles of one triangle are congruent to corresponding angles of another triangle.

Why
Then we know that the two triangles are similar by the AAA property.

Solution
Since vertical angles are congruent,

$$\angle PQR \cong \angle NQM$$

In the figure, we can view $\overleftrightarrow{PN}$ as a transversal cutting parallel line segments $\overline{PR}$ and $\overline{MN}$. Since alternate interior angles are then congruent, we have:

$$\angle RPQ \cong \angle MNQ$$

Furthermore, we can view $\overleftrightarrow{RM}$ as a transversal cutting parallel line segments $\overline{PR}$ and $\overline{MN}$. Since alternate interior angles are then congruent, we have:

$$\angle QRP \cong \angle QMN$$

These observations are summarized in the figure below.

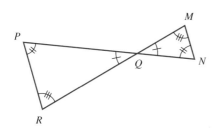

In the figure, we see that corresponding angles of $\triangle PQR$ are congruent to corresponding angles of $\triangle NQM$. By the AAA similarity theorem, we can conclude that

$$\triangle PQR \sim \triangle NQM$$

Self Check 5
In the figure below, $\overline{YA} \parallel \overline{ZB}$. Are $\triangle XYA$ and $\triangle XZB$ similar triangles?

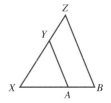

Answer: yes, by the AAA similarity theorem; $\angle X \cong \angle X$, $\angle XYA \cong \angle XZB$, $\angle XAY \cong \angle XBZ$

Now Try
Problems 41 and 43

EXAMPLE 6 In the figure on the right, $\triangle RST \sim \triangle JKL$. Find **a.** *x* and **b.** *y*.

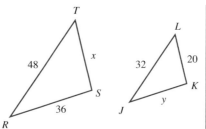

Strategy

To find *x*, we will write a proportion of corresponding sides so that *x* is the only unknown. Then we will solve the proportion for *x*. We will use a similar method to find *y*.

Why

Since $\triangle RST \sim \triangle JKL$, we know that the lengths of corresponding sides of $\triangle RST$ and $\triangle JKL$ are proportional.

Solution

a. When we write $\triangle RST \sim \triangle JKL$, a correspondence between the vertices of the two triangles is established.

$$\triangle RST \sim \triangle JKL$$

The lengths of corresponding sides of these similar triangles are proportional.

$\dfrac{RT}{JL} = \dfrac{ST}{KL}$ Each fraction is a ratio of a side length of $\triangle RST$ to its corresponding side length of $\triangle JKL$.

$\dfrac{48}{32} = \dfrac{x}{20}$ Substitute: $RT = 48$, $JL = 32$, $ST = x$, and $KL = 20$.

$48(20) = 32x$ Find each cross product and set them equal.

$960 = 32x$ Do the multiplication.

$30 = x$ To isolate *x*, undo the multiplication by 32 by dividing both sides by 32.

$x = 30$

b. To find *y*, we write a proportion of corresponding side lengths in such a way that *y* is the only unknown.

$\dfrac{RT}{JL} = \dfrac{RS}{JK}$

$\dfrac{48}{32} = \dfrac{36}{y}$ Substitute: $RT = 48$, $JL = 32$, $RS = 36$, and $JK = y$.

$48y = 32(36)$ Find each cross product and set them equal.

$48y = 1,152$ Do the multiplication.

$y = 24$ To isolate *y*, undo the multiplication by 48 by dividing both sides by 48.

Self Check 6

In the figure below, $\triangle DEF \sim \triangle GHI$. Find **a.** *x* and **b.** *y*.

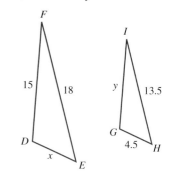

Answers: **a.** 6   **b.** 11.25

Now Try

Problem 53

4. Use similar triangles to find unknown lengths in application problems.

Similar triangles and proportions can be used to find lengths that would normally be difficult to measure. For example, we can use the reflective properties of a mirror to calculate the height of a flagpole while standing safely on the ground.

EXAMPLE 7 To determine the height of a flagpole, a woman walks to a point 20 feet from its base, as shown on the next page. Then she takes a mirror from her purse, places it on the ground, and walks 2 feet farther away, where she can see the top of the pole reflected in the mirror. Find the height of the pole.

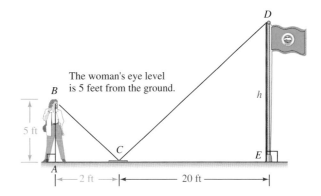

The woman's eye level is 5 feet from the ground.

Strategy

We will show that $\triangle ABC \sim \triangle EDC$.

Why

Then we can write a proportion of corresponding sides so that h is the only unknown and we can solve the proportion for h.

Solution

To show that $\triangle ABC \sim \triangle EDC$, we begin by applying an important fact about mirrors. When a beam of light strikes a mirror, it is reflected at the same angle as it hits the mirror. Therefore, $\angle BCA \cong \angle DCE$. Furthermore, $\angle A \cong \angle E$ because the woman and the flagpole are perpendicular to the ground. Finally, if two pairs of corresponding angles are congruent, it follows that the third pair of corresponding angles are also congruent: $\angle B \cong \angle D$. By the AAA similarity theorem, we conclude that $\triangle ABC \sim \triangle EDC$.

Since the triangles are similar, the lengths of their corresponding sides are in proportion. If we let h represent the height of the flagpole, we can find h by solving the following proportion.

$$\frac{h}{5} = \frac{20}{2}$$

$2h = 5(20)$ Find each cross product and set them equal.

$2h = 100$

$h = 50$

The flagpole is 50 feet tall.

Now Try

Problem 83

STUDY SET   Section 8

VOCABULARY   *Fill in the blanks.*

1. _____ triangles are the same size and the same shape.

2. When we match the vertices of $\triangle ABC$ with the vertices of $\triangle DEF$, as shown below, we call this matching of points a _____.

$A \leftrightarrow D \quad B \leftrightarrow E \quad C \leftrightarrow F$

3. Two angles or two line segments with the same measure are said to be _____.

4. Corresponding _____ of congruent triangles are congruent.

5. If two triangles are _____, they have the same shape but not necessarily the same size.

6. A mathematical statement that two ratios (fractions) are equal, such as $\frac{x}{18} = \frac{4}{9}$, is called a _____.

CONCEPTS

7. Refer to the triangles in the figure below.

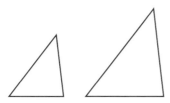

a. Do these triangles appear to be congruent? Explain why or why not.

b. Do these triangles appear to be similar? Explain why or why not.

8. a. Draw a triangle that is congruent to △*CDE* shown below. Label it △*ABC*.

b. Draw a triangle that is similar to, but not congruent to, △*CDE*. Label it △*MNO*.

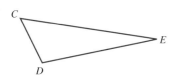

Fill in the blanks.

9. $\triangle XYZ \cong \triangle$ _____

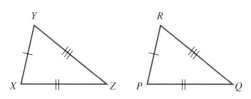

10. $\triangle$ _____ $\cong \triangle DEF$

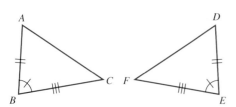

11. $\triangle RST \sim \triangle$ _____

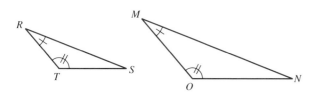

12. $\triangle$ _____ $\sim \triangle TAC$

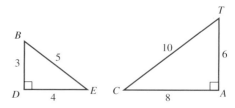

13. Name the six corresponding parts of the congruent triangles shown below.

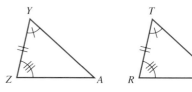

14. Name the six corresponding parts of the congruent triangles shown below.

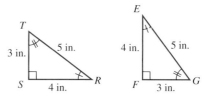

Fill in the blanks.

15. Two triangles are _____ if and only if their vertices can be matched so that the corresponding sides and the corresponding angles are congruent.

16. SSS property: If three _____ of one triangle are congruent to three _____ of a second triangle, the triangles are congruent.

17. SAS property: If two sides and the _____ between them in one triangle are congruent, respectively, to two sides and the _____ between them in a second triangle, the triangles are congruent.

18. ASA property: If two angles and the _____ between them in one triangle are congruent, respectively, to two angles and the _____ between them in a second triangle, the triangles are congruent.

Solve each proportion.

19. $\frac{x}{15} = \frac{20}{3}$

20. $\frac{5}{8} = \frac{35}{x}$

21. $\frac{h}{2.6} = \frac{27}{13}$

22. $\frac{11.2}{4} = \frac{h}{6}$

Fill in the blanks.

23. Two triangles are similar if and only if their vertices can be matched so that corresponding angles are congruent and the lengths of corresponding sides are _____.

24. If the angles of one triangle are congruent to corresponding angles of another triangle, the triangles are _____.

25. Congruent triangles are always similar, but similar triangles are not always _____.

26. For certain application problems, similar triangles and _____ can be used to find lengths that would normally be difficult to measure.

NOTATION *Fill in the blanks.*

27. The symbol ≅ is read as "____ ____ ____."

28. The symbol ~ is read as "____ ____ ____."

29. Use markings to show the congruent parts of the triangles shown below.

$$\angle K \cong \angle H \qquad \overline{KR} \cong \overline{HJ} \qquad \angle M \cong \angle E$$

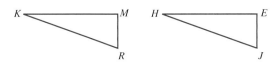

30. Use markings to show the congruent parts of the triangles shown below.

$$\angle P \cong \angle T \qquad \overline{LP} \cong \overline{RT} \qquad \overline{FP} \cong \overline{ST}$$

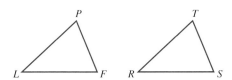

GUIDED PRACTICE

Name the six corresponding parts of the congruent triangles. **See Objective 1.**

31. $\overline{AC} \cong$ _____

$\overline{DE} \cong$ _____

$\overline{BC} \cong$ _____

$\angle A \cong$ _____

$\angle E \cong$ _____

$\angle F \cong$ _____

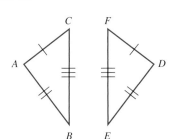

32. $\overline{AB} \cong$ _____

$\overline{EC} \cong$ _____

$\overline{AC} \cong$ _____

$\angle D \cong$ _____

$\angle B \cong$ _____

$\angle 1 \cong$ _____

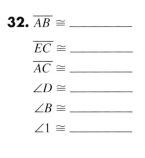

33. Refer to the figure below, where △BCD ≅ △MNO.

  a. Name the six congruent corresponding parts of the triangles. **See Example 1.**

  b. Find m(∠N).

  c. Find m($\overline{MO}$).

  d. Find m($\overline{CD}$).

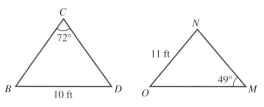

34. Refer to the figure below, where △DCG ≅ △RST.

  a. Name the six congruent corresponding parts of the triangles. **See Example 1.**

  b. Find m(∠R).

  c. Find m($\overline{DG}$).

  d. Find m($\overline{ST}$).

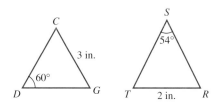

Determine whether each pair of triangles is congruent. If they are, tell why. **See Examples 2 and 3.**

35.

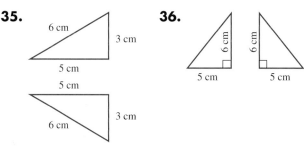

36.

37.

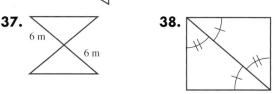

38.

39. Refer to the similar triangles shown below. **See Example 4.**

  a. Name 3 pairs of congruent angles.

  b. Complete each proportion.

$$\frac{LM}{HJ} = \frac{}{} \qquad \frac{MR}{JE} = \frac{}{HE} \qquad \frac{}{HJ} = \frac{LR}{HE}$$

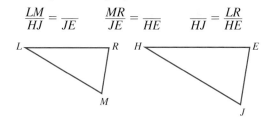

40. Refer to the similar triangles shown below.
See Example 4.

   a. Name 3 pairs of congruent angles.

   b. Complete each proportion.

$$\frac{WY}{DF} = \frac{}{FE} \qquad \frac{WX}{} = \frac{YX}{FE} \qquad \frac{}{EF} = \frac{WY}{DF}$$

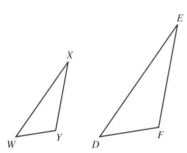

Tell whether the triangles are similar. **See Example 5.**

41.

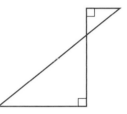

42.

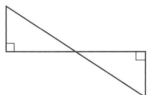

43.

44.

45.

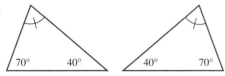

46.

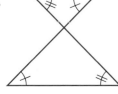

47.

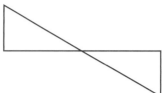

48.

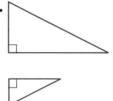

49. $\overline{XY} \parallel \overline{ZD}$

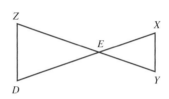

50.

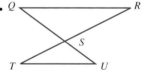

51.

52.

$\triangle MSN \sim \triangle TPR.$ *Find x and y.* **See Example 6.**

53.

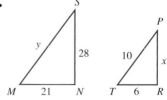

54.

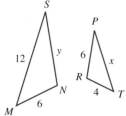

△MSN ~ △TPN. Find x and y. ***See Example 6.***

55.

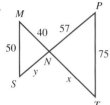

56.

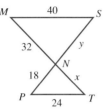

TRY IT YOURSELF

Tell whether each statement is true. If a statement is false, tell why.

57. If three sides of one triangle are the same length as the corresponding three sides of a second triangle, the triangles are congruent.

58. If two sides of one triangle are the same length as two sides of a second triangle, the triangles are congruent.

59. If two sides and an angle of one triangle are congruent, respectively, to two sides and an angle of a second triangle, the triangles are congruent.

60. If two angles and the side between them in one triangle are congruent, respectively, to two angles and the side between them in a second triangle, the triangles are congruent.

Determine whether each pair of triangles are congruent. If they are, tell why.

61.

62.

63.

64.

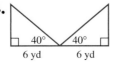

65. $\overline{AB} \parallel \overline{DE}$

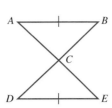

66. $\overline{XY} \parallel \overline{ZQ}$

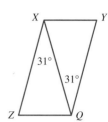

67.

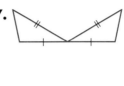

68.

In Problems 69 and 70, △ABC ≅ △DEF. Find x and y.

69.

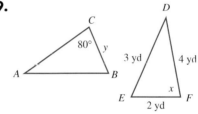

70.

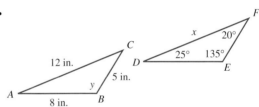

In Problems 71 and 72, find x and y.

71. △ABC ≅ △ABD

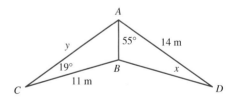

72. △ABC ≅ △DEC

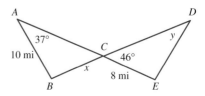

In Problems 73–76, find x.

73.
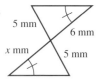

74.
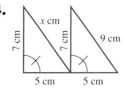

75.
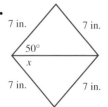

76.
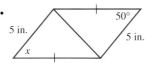

77. If $\overline{DE}$ in the figure below is parallel to $\overline{AB}$, $\triangle ABC$ will be similar to $\triangle DEC$. Find x.

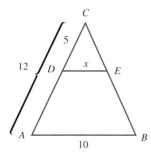

78. If $\overline{SU}$ in the figure below is parallel to $\overline{TV}$, $\triangle SRU$ will be similar to $\triangle TRV$. Find x.

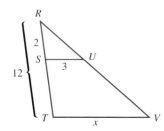

79. If $\overline{DE}$ in the figure below is parallel to $\overline{CB}$, $\triangle EAD$ will be similar to $\triangle BAC$. Find x.

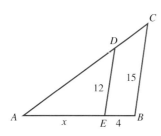

80. If $\overline{HK}$ in the figure below is parallel to $\overline{AB}$, $\triangle HCK$ will be similar to $\triangle ACB$. Find x.

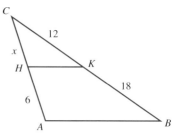

APPLICATIONS

81. SEWING   The pattern that is sewn on the rear pocket of a pair of blue jeans is shown below. If $\triangle AOB \cong \triangle COD$, how long is the stitching from point A to point D?

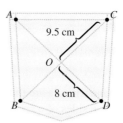

82. CAMPING   The base of the tent pole is placed at the midpoint between the stake at point A and the stake at point B, and it is perpendicular to the ground, as shown below, Explain why $\triangle ACD \cong \triangle BCD$.

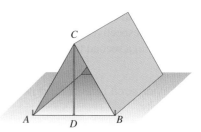

83. HEIGHT OF A TREE   The tree shown below casts a shadow 24 feet long when a man 6 feet tall casts a shadow 4 feet long. Find the height of the tree.

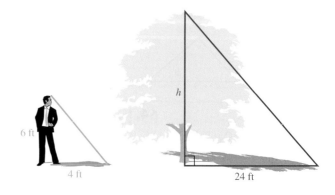

84. HEIGHT OF A BUILDING   A man places a mirror on the ground and sees the reflection of the top of a building, as shown below. Find the height of the building.

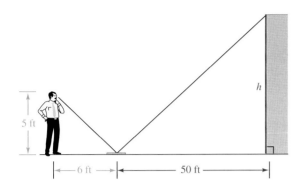

85. WIDTH OF A RIVER   Use the dimensions in the figure below to find *w*, the width of the river.

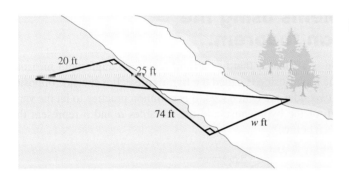

86. FLIGHT PATH   An airplane ascends 200 feet as it flies a horizontal distance of 1,000 feet, as shown in the figure. How much altitude is gained as it flies a horizontal distance of 1 mile? (*Hint:* 1 mile = 5,280 feet.)

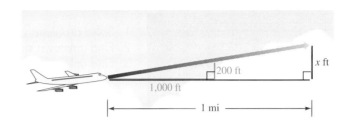

WRITING

87. Tell whether the statement is true or false. Explain your answer.

a. Congruent triangles are always similar.

b. Similar triangles are always congruent.

88. Explain why there is no SSA property for congruent triangles.

9 *The Pythagorean Theorem and Special Triangles*

Objectives

1. Solve problems using the Pythagorean theorem.
2. Find unknown side lengths of 45°–45°–90° triangles.
3. Find unknown side lengths of 30°–60°–90° triangles.

PYTHAGORAS

A **theorem** is a mathematical statement that can be proven. In this section, we will discuss one of the most widely used theorems of geometry—the Pythagorean theorem. It is named after Pythagoras, a Greek mathematician who lived about 2,500 years ago. He is thought to have been the first to develop a proof of it. The Pythagorean theorem expresses the relationship between the lengths of the sides of any right triangle. We will also study two specific types of right triangles and we will derive some formulas that can be used to find their missing side lengths.

1. Solve problems using the Pythagorean theorem.

Recall that a right triangle is a triangle that has a right angle (an angle with measure 90°). In a right triangle, the longest side is called the **hypotenuse.** It is the side opposite the right angle. The other two sides are called **legs.** It is common practice to let the variable c represent the length of the hypotenuse and the variables a and b represent the lengths of the legs, as shown below.

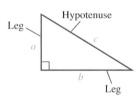

If we know the lengths of any two sides of a right triangle, we can find the length of the third side using the **Pythagorean theorem.**

Pythagorean theorem

> If a and b are the lengths of two legs of a right triangle and c is the length of the hypotenuse, then
> $$a^2 + b^2 = c^2$$

In words, the Pythagorean theorem is expressed as follows:

> *In a right triangle, the sum of the squares of the lengths of the two legs is equal to the square of the length of the hypotenuse.*

CAUTION   When using the **Pythagorean equation** $a^2 + b^2 = c^2$, we can let a represent the length of either leg of the right triangle. We then let b represent the length of the other leg. The variable c must always represent the length of the hypotenuse.

EXAMPLE 1   Find the length of the hypotenuse of the right triangle shown on the right.

3 in.

4 in.

Strategy
We will use the Pythagorean theorem to find the length of the hypotenuse.

Why
If we know the lengths of any two sides of a right triangle, we can find the length of the third side using the Pythagorean theorem.

Solution
We will let $a = 3$ and $b = 4$, and substitute into the Pythagorean equation to find c.

$a^2 + b^2 = c^2$ This is the Pythagorean equation.

$3^2 + 4^2 = c^2$ Substitute 3 for a and 4 for b.

$9 + 16 = c^2$ Evaluate each exponential expression.

$25 = c^2$ Do the addition.

$c^2 = 25$ Reverse the sides of the equation so that c^2 is on the left.

$a = 3$ in.

c

$b = 4$ in.

According to the square root property, the equation $c^2 = 25$ has two solutions: $c = \sqrt{25}$ and $c = -\sqrt{25}$. Since c represents the length of a side of a triangle, it follows that c is the positive square root of 25.

$c = \sqrt{25}$

$c = 5$

The length of the hypotenuse is 5 in.

Self Check 1
Find the length of the hypotenuse of the right triangle shown below.

5 ft

12 ft

Answer: 13 ft

Now Try
Problem 23

EXAMPLE 2   *Firefighting.*   To fight a forest fire, the forestry department plans to clear a rectangular fire break around the fire, as shown in the following figure. Crews are equipped with mobile communications that have a 3,000-yard range. Can crews at points A and B remain in radio contact?

Strategy
We will use the Pythagorean theorem to find the distance between points A and B.

Why
If the distance is less than 3,000 yards, the crews can communicate by radio. If it is greater than 3,000 yards, they cannot.

Solution
The line segments connecting points A, B, and C form a right triangle. To find the distance c from point A to point B, we can use the Pythagorean equation, substituting 2,400 for a and 1,000 for b and solving for c.

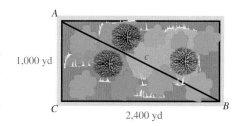

A

1,000 yd

c

C

2,400 yd

B

$a^2 + b^2 = c^2$ This is the Pythagorean equation.

$2{,}400^2 + 1{,}000^2 = c^2$ Substitute for a and b.

$5{,}760{,}000 + 1{,}000{,}000 = c^2$ Evaluate each exponential expression.

$6{,}760{,}000 = c^2$ Do the addition.

$c^2 = 6{,}760{,}000$ Reverse the sides of the equation so that c^2 is on the left.

Self Check 2
In Example 2, can the crews communicate by radio if the distance from point B to point C remains the same but the distance from point A to point C increases to 2,520 yards?

$$c = \sqrt{6{,}760{,}000}$$ If $c^2 = 6{,}760{,}000$, then c must be a square root of 6,760,000. Because c represents a length, it must be the positive square root of 6,760,000.

$$c = 2{,}600$$ Use a calculator to find the square root.

The two crews are 2,600 yards apart. Because this distance is less than the 3,000-yard range of the radios, they can communicate by radio.

Answer: no

Now Try
Problems 29 and 87

EXAMPLE 3   The lengths of two sides of a right triangle are given in the figure on the right. Find the missing side length.

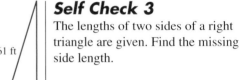

Strategy
We will use the Pythagorean theorem to find the missing side length.

Why
If we know the lengths of any two sides of a right triangle, we can find the length of the third side using the Pythagorean theorem.

Solution
We may substitute 11 for either a or b, but 61 must be substituted for the length c of the hypotenuse. If we choose to substitute 11 for b, we can find the unknown side length a as follows.

$$a^2 + b^2 = c^2$$ This is the Pythagorean equation.

$$a^2 + 11^2 = 61^2$$ Substitute 11 for b and 61 for c.

$$a^2 + 121 = 3{,}721$$ Evaluate each exponential expression.

$$a^2 + 121 - 121 = 3{,}721 - 121$$ To isolate a^2 on the left side, undo the addition of 121 by subtracting 121 from both sides.

$$a^2 = 3{,}600$$ Do the subtractions.

$$a = \sqrt{3{,}600}$$ If $a^2 = 3{,}600$, then a must be a square root of 3,600. Because a represents a length, it must be the positive square root of 3,600.

$$a = 60$$

The missing side length is 60 ft.

Self Check 3
The lengths of two sides of a right triangle are given. Find the missing side length.

Answer: 56 in.

Now Try
Problem 31

EXAMPLE 4   Refer to the triangle on the right. Find the missing side length. Give the exact answer and an approximation to the nearest hundredth.

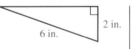

Strategy
We will use the Pythagorean theorem to find the missing side length.

Why
If we know the lengths of any two sides of a right triangle, we can find the length of the third side using the Pythagorean theorem.

Solution
We may substitute 2 for either a or b, but 6 must be substituted for the length c of the hypotenuse. If we choose to substitute 2 for a, we can find the unknown side length b as follows.

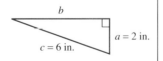

Self Check 4
Refer to the triangle below. Find the missing side length. Give the exact answer and an approximation to the nearest tenth.

$a^2 + b^2 = c^2$ This is the Pythagorean equation

$2^2 + b^2 = 6^2$ Substitute 2 for a and 6 for c.

$4 + b^2 = 36$ Evaluate each exponential expression.

$4 + b^2 - 4 = 36 - 4$ To isolate b^2 on the left side, undo the addition of 4 by subtracting 4 from both sides.

$b^2 = 32$

Now we apply the square root property. Since b represents the length of a side of a triangle, we consider only the positive square root.

$b = \sqrt{32}$

$b = 4\sqrt{2}$ Simplify the radical expression: $\sqrt{32} = \sqrt{16 \cdot 2} = 4\sqrt{2}$.

The missing side length is exactly $4\sqrt{2}$ inches long. We can use a calculator to approximate the length. To the nearest tenth, it is 5.7 inches.

Answer: $\sqrt{24}$ m $= 2\sqrt{6}$ m ≈ 4.9 m

Now Try
Problem 39

For more complicated problems involving right triangles, it is often helpful to use the following five-step problem-solving strategy.

EXAMPLE 5   The longer leg of a right triangle is 3 units longer than the shorter leg. If the hypotenuse is 6 units longer than the shorter leg, find the lengths of the sides of the triangle.

Analyze the problem   We begin by drawing a right triangle and labeling the legs and the hypotenuse.

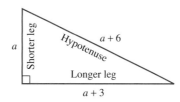

Form an equation   Let $a =$ length of the shorter leg. Then the length of the hypotenuse is $a + 6$ and the length of the longer leg is $a + 3$. By the Pythagorean theorem, we have

$$\left(\begin{array}{c}\text{the length of}\\\text{the shorter leg}\end{array}\right)^2 \quad \text{plus} \quad \left(\begin{array}{c}\text{the length of}\\\text{the longer leg}\end{array}\right)^2 \quad \text{equals} \quad \left(\begin{array}{c}\text{the length of the}\\\text{hypotenuse.}\end{array}\right)^2$$

$$a^2 \qquad + \qquad (a+3)^2 \qquad = \qquad (a+6)^2$$

Solve the equation

$a^2 + (a + 3)^2 = (a + 6)^2$

$a^2 + a^2 + 6a + 9 = a^2 + 12a + 36$ Find $(a + 3)^2$ and $(a + 6)^2$.

$2a^2 + 6a + 9 = a^2 + 12a + 36$ On the left side: $a^2 + a^2 = 2a^2$.

$a^2 - 6a - 27 = 0$ To get 0 on the right side, subtract a^2, $12a$, and 36 from both sides. Note that this is a quadratic equation.

$(a - 9)(a + 3) = 0$ Factor the trinomial, $a^2 - 6a - 27$.

$a - 9 = 0$ or $a + 3 = 0$ Set each factor equal to 0.

$a = 9$ $\qquad\qquad$ $a = -3$ Solve each equation.

State the conclusion   Since a side cannot have a negative length, we discard the solution -3. Thus, the shorter leg is 9 units long, the hypotenuse is $9 + 6 = 15$ units long, and the longer leg is $9 + 3 = 12$ units long.

Check the result   The longer leg, 12, is 3 units longer than the shorter leg, 9. The hypotenuse, 15, is 6 units longer than the shorter leg, 9, and the side lengths satisfy the Pythagorean equation. So the results check.

$$9^2 + 12^2 \overset{?}{=} 15^2$$
$$81 + 144 \overset{?}{=} 225$$
$$225 = 225$$

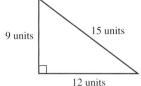

Now Try
Problem 49

Recall that if a mathematical statement is written in the form *if p . . . , then q . . . ,* we call the statement *if q . . . , then p . . .* its **converse.** The converses of some statements are true, while the converses of other statements are false. It is interesting to note that the converse of the Pythagorean theorem is true.

Converse of the Pythagorean theorem	If a triangle has three sides of lengths a, b, and c, such that $a^2 + b^2 = c^2$, then the triangle is a right triangle.

EXAMPLE 6   Is the triangle shown on the right a right triangle?

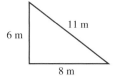

Strategy
We will substitute the side lengths, 6, 8, and 11, into the Pythagorean equation $a^2 + b^2 = c^2$.

Why
By the converse of the Pythagorean theorem, the triangle is a right triangle if a true statement results. The triangle is not a right triangle if a false statement results.

Solution
We must substitute the longest side length, 11, for c, because it is the possible hypotenuse. The lengths of 6 and 8 may be substituted for either a or b.

$a^2 + b^2 = c^2$ This is the Pythagorean equation.

$6^2 + 8^2 \overset{?}{=} 11^2$ Substitute 6 for a, 8 for b, and 11 for c.

$36 + 64 \overset{?}{=} 121$ Evaluate each exponential expression.

$100 = 121$ This is a false statement.

Since $100 \neq 121$, the triangle is not a right triangle.

Self Check 6
Is the triangle below a right triangle?

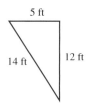

Answer: no

Now Try
Problem 51

2. Find unknown side lengths of 45°–45°–90° triangles.

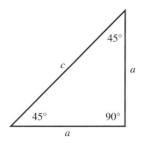

An **isosceles right triangle** is a right triangle with two legs of equal length. Isosceles right triangles have angle measures of 45°, 45°, and 90°. If we know the length of one leg of an isosceles right triangle, we can use the Pythagorean theorem to find the length of the hypotenuse. Since the triangle shown in the margin is a right triangle, we have

$$a^2 + b^2 = c^2 \qquad \text{This is the Pythagorean equation.}$$

$$a^2 + a^2 = c^2 \qquad \text{Both legs are } a \text{ units long, so replace } b \text{ with } a.$$

$$2a^2 = c^2 \qquad \text{On the left side, combine like terms.}$$

$$c^2 = 2a^2 \qquad \text{Reverse the sides of the equation so that } c^2 \text{ is on the left.}$$

$$c = \sqrt{2a^2} \qquad \text{If } c^2 = 2a^2, \text{ then } c \text{ must be a square root of } 2a^2. \text{ Because } c \text{ represents a length, it must be the positive square root of } 2a^2.$$

$$c = a\sqrt{2} \qquad \text{Simplify the radical: } \sqrt{2a^2} = \sqrt{2}\sqrt{a^2} = \sqrt{2}a = a\sqrt{2}.$$

This result is a formula that can be used to find the length of the hypotenuse of an isosceles right triangle, given the length of a leg.

Isosceles right triangles

> The length c of the hypotenuse of an isosceles right triangle is $\sqrt{2}$ times the length a of either leg:
>
> $$c = a\sqrt{2}$$

! **CAUTION** The formula $c = a\sqrt{2}$ may also be written as $c = \sqrt{2}a$. However, we must be very careful when writing it that way, because only the 2 (*not* the a) is under the radical symbol.

EXAMPLE 7 One leg of an isosceles right triangle is 10 feet long. Find the exact length of the hypotenuse and an approximation to the nearest hundredth.

Strategy
We will find the length of the hypotenuse by substituting 10 for a in the formula $c = a\sqrt{2}$.

Why
The variable c represents the unknown length of the hypotenuse.

Solution

$$c = a\sqrt{2} \qquad \text{This is the formula for the length of the hypotenuse of an isosceles right triangle.}$$

$$c = 10\sqrt{2} \qquad \text{Substitute 10 for } a, \text{ the length of one leg.}$$

The exact length of the hypotenuse is $10\sqrt{2}$ feet. To the nearest hundredth, the length of the hypotenuse is 14.14 feet.

Self Check 7
One leg of an isosceles right triangle is 12 meters long. Find the exact length of the hypotenuse and an approximation to the nearest hundredth.

Answer: $12\sqrt{2}$ m ≈ 16.97 m

Now Try
Problem 55

EXAMPLE 8 Find the exact length and an approximation to the nearest hundredth of the length of each leg of the triangle shown on the right.

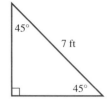

Strategy

We will find the length of each leg of the triangle by substituting 25 for c in the formula $c = a\sqrt{2}$ and solving for a.

Why

The variable a represents the unknown length of each leg of the triangle.

Solution

$$c = a\sqrt{2}$$ This is the formula for the length of the hypotenuse of an isosceles right triangle.

$$25 = a\sqrt{2}$$ Substitute 25 for c, the length of the hypotenuse.

$$\frac{25}{\sqrt{2}} = \frac{a\sqrt{2}}{\sqrt{2}}$$ To isolate a, undo the multiplication by $\sqrt{2}$ by dividing both sides by $\sqrt{2}$.

$$\frac{25}{\sqrt{2}} = a$$ Simplify the right side.

$$a = \frac{25}{\sqrt{2}}$$ Reverse the sides of the equation so that a is on the left.

$$a = \frac{25}{\sqrt{2}} \cdot \frac{\sqrt{2}}{\sqrt{2}}$$ Rationalize the denominator.

$$a = \frac{25\sqrt{2}}{2}$$ Simplify the denominator: $\sqrt{2} \cdot \sqrt{2} = 2$.

The exact length of each leg of the triangle is $\frac{25\sqrt{2}}{2}$ cm. We can use a calculator to approximate this. To the nearest hundredth, the length of each leg is 17.68 cm.

Self Check 8

Find the exact length and an approximation to the nearest hundredth of the length of each leg of the triangle shown below.

Answer: $\frac{7\sqrt{2}}{2}$ ft ≈ 4.95 ft

Now Try

Problem 59

3. Find unknown side lengths of 30°–60°–90° triangles.

Recall that an equilateral triangle is a triangle with three sides of equal length and three 60° angles. Each side of the equilateral triangle shown below is $2a$ units long. If an altitude (shown in green) is drawn to its base, the altitude bisects the base and divides the equilateral triangle into two congruent 30°–60°–90° triangles. We see that the shorter leg of each 30°–60°–90° triangle (the side *opposite* the 30° angle) is a units long; that is, it is half as long as the hypotenuse. Another way to express this relationship is that the hypotenuse is twice as long as the shorter leg.

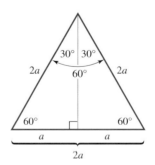

 CAUTION   In a 30°–60°–90° triangle, the shorter leg is opposite the 30° angle, and the longer leg is opposite the 60° angle.

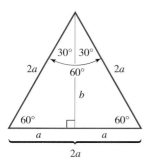

We can discover another important relationship between the legs of a 30°–60°–90° triangle if we find the length b of the altitude shown in the margin. We begin by applying the Pythagorean theorem to one of the 30°–60°–90° triangles.

$a^2 + b^2 = c^2$ This is the Pythagorean equation.

$a^2 + b^2 = (2a)^2$ Since the hypotenuse is $2a$ units long, replace c with $2a$.

$a^2 + b^2 = 4a^2$ Find the 2nd power of $2a$: $(2a)^2 = (2a)(2a) = 4a^2$.

$b^2 = 3a^2$ To isolate b^2, subtract a^2 from both sides: $4a^2 - a^2 = 3a^2$.

$b = \sqrt{3a^2}$ Apply the square root property. Since b represents the length of a side of a triangle, we discard the solution $b = -\sqrt{3a^2}$.

$b = a\sqrt{3}$ Simplify the radical: $\sqrt{3a^2} = \sqrt{3}\,\sqrt{a^2} = \sqrt{3}\,a = a\sqrt{3}$.

We see that the longer leg of the 30°–60°–90° triangle is $\sqrt{3}$ times as long as the shorter leg.

 CAUTION   The formula $b = a\sqrt{3}$ may also be written as $b = \sqrt{3}a$. However, we must be very careful when writing it that way, because only the 3 (not the a) is under the radical symbol.

30°–60°–90° triangles

> For any 30°–60°–90° triangle,
>
> **1.** The length c of the hypotenuse is twice the length a of the shorter leg: $c = 2a$
>
> **2.** The length b of the longer leg is $\sqrt{3}$ times the length a of the shorter leg: $b = a\sqrt{3}$

EXAMPLE 9   Find the length of the hypotenuse and the longer leg of the right triangle shown on the right. Approximate any exact answers that contain a radical by rounding to the nearest hundredth.

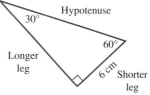

Strategy
We will find the length of the hypotenuse by substituting 6 for a in the formula $c = 2a$ and evaluating the right side. We will find the length of the longer leg by substituting 6 for a in the formula $b = a\sqrt{3}$.

Why
The variable c represents the unknown length of the hypotenuse and the variable b represents the unknown length of the longer leg.

Solution
To find the length of the hypotenuse, we proceed as follows:

$c = 2a$ This is the formula for the length of the hypotenuse of a
30°–60°–90° triangle.

$c = 2(6)$ Substitute 6 for a, the length of the shorter leg.

$c = 12$ Do the multiplication.

The hypotenuse is 12 centimeters long.
   To find the length of the longer leg, we proceed as follows:

$b = a\sqrt{3}$ This is the formula for the length of the longer leg of a
30°–60°–90° triangle.

$b = 6\sqrt{3}$ Substitute 6 for a, the length of the shorter leg.

The longer leg is exactly $6\sqrt{3}$ centimeters long. We can use a calculator to approximate this. To the nearest hundredth, the length of the longer leg is 10.39 centimeters.

Self Check 9
Find the length of the hypotenuse and the longer leg of a 30°–60°–90° triangle if the shorter leg is 8 centimeters long. Approximate any exact answers that contain a radical by rounding to the nearest hundredth.

Answer: 16 cm,
$8\sqrt{3}$ cm ≈ 13.86 cm

Now Try
Problem 63

EXAMPLE 10   Find the missing side lengths of the triangle shown on the right. Approximate any exact answers that contain a radical by rounding to the nearest hundredth.

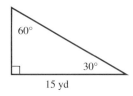

Strategy

We will find the length of the shorter leg first.

Why

Once we know the length of the shorter leg, we can multiply it by 2 to find the length of the hypotenuse.

Solution

From the figure, the length of the longer leg is 15 yards. We can find the length a of the shorter leg as follows:

$b = a\sqrt{3}$	This is the formula for the length of the longer leg of a $30°$–$60°$–$90°$ triangle.
$15 = a\sqrt{3}$	The longer leg is 15 yards long.
$\dfrac{15}{\sqrt{3}} = \dfrac{a\sqrt{3}}{\sqrt{3}}$	To isolate a, undo the multiplication by $\sqrt{3}$ by dividing both sides by $\sqrt{3}$.
$\dfrac{15}{\sqrt{3}} = a$	Simplify the right side.
$a = \dfrac{15}{\sqrt{3}}$	Reverse the sides of the equation so that a is on the left.
$a = \dfrac{15}{\sqrt{3}} \cdot \dfrac{\sqrt{3}}{\sqrt{3}}$	Rationalize the denominator.
$a = \dfrac{15\sqrt{3}}{3}$	Simplify the denominator: $\sqrt{3} \cdot \sqrt{3} = 3$.
$a = 5\sqrt{3}$	Simplify the fraction: $\dfrac{15}{3} = 5$.

The length of the shorter leg is exactly $5\sqrt{3}$ yards. We can use a calculator to approximate this. This is about 8.66 yards.

Since the length of the hypotenuse is twice as long as the shorter leg, the length of the hypotenuse is $2 \cdot 5\sqrt{3}$ yards, or $10\sqrt{3}$ yards. To the nearest hundredth, the hypotenuse is 17.32 yards long.

Self Check 10

Find the missing side lengths of the triangle shown below. Approximate any exact answers that contain a radical by rounding to the nearest hundredth.

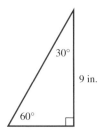

Answer: $3\sqrt{3}$ in. ≈ 5.20 in., $6\sqrt{3}$ in. ≈ 10.39 in.

Now Try

Problem 67

EXAMPLE 11   ***Stretching exercises.*** A doctor prescribed the exercise shown in figure (a). The patient was instructed to raise his leg to an angle of 60° and hold the position for 10 seconds. If the patient's leg is 36 inches long, how high off the floor will his foot be when his leg is held at the proper angle?

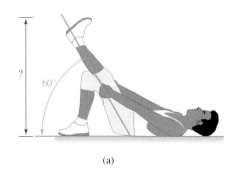

(a)

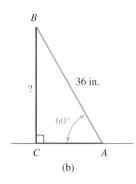

(b)

Strategy

This situation is modeled by a $30°$–$60°$–$90°$ right triangle. We will begin by finding the length of the shorter leg.

Why

Once we know the length of the shorter leg, we can find the length of the longer leg, which represents the distance the patient's foot is off the ground.

Solution

In figure (b), we see that a $30°–60°–90°$ triangle, which we will call $\triangle ABC$, models the situation. Since the side opposite the $30°$ angle of a $30°–60°–90°$ triangle is half as long as the hypotenuse, side $\overline{CA}$ is 18 inches long.

Since the length of the side opposite the $60°$ angle is the length of the side opposite the $30°$ angle times $\sqrt{3}$, side $\overline{BC}$ is $18\sqrt{3}$, or about 31 inches long. So the patient's foot will be about 31 inches from the floor when his leg is in the proper stretching position.

Now Try

Problem 71

STUDY SET   Section 9

VOCABULARY   *Fill in the blanks.*

1. In a right triangle, the side opposite the $90°$ angle is called the _____. The other two sides are called _____.

2. The Pythagorean theorem is named after the Greek mathematician, _____, who is thought to have been the first to prove it.

3. The _____ theorem states that in any right triangle, the square of the length of the hypotenuse is equal to the sum of the squares of the lengths of the two legs.

4. $a^2 + b^2 = c^2$ is called the Pythagorean _____.

5. An _____ right triangle is a right triangle with two legs of equal length.

6. An _____ triangle has three sides of equal length and three $60°$ angles.

CONCEPTS   *Fill in the blanks.*

7. If a and b are the lengths of two legs of a right triangle and c is the length of the hypotenuse, then $+$ $=$.

8. In an isosceles right triangle, the length of the hypotenuse is the length of one leg times   .

9. In a $30°–60°–90°$ triangle, the shorter leg is opposite the $\_\_°$ angle, and the longer leg is opposite the $\_\_°$ angle.

10. In a $30°–60°–90°$ triangle:

a. The hypotenuse is _____ as long as the shorter leg. Or put another way, the shorter leg is _____ as long as the hypotenuse.

b. The length of the longer leg of a $30°–60°–90°$ triangle is the length of the shorter leg times   .

11. The two solutions of $c^2 = 36$ are $c =$ or $c =$. If c represents the length of the hypotenuse of a right triangle, then we can discard the solution   .

12. The converse of the Pythagorean theorem: If a triangle has three sides of lengths a, b, and c, such that $a^2 + b^2 = c^2$, then the triangle is a _____ triangle.

13. Use a protractor to draw an example of each type of triangle.

a. A right triangle

b. A right isosceles triangle

c. A $30°–60°–90°$ triangle

14. Refer to the triangle on the right.

a. What side is the hypotenuse?

b. What side is the longer leg?

c. What side is the shorter leg?

15. What is the first step when solving the equation $25 + b^2 = 81$ for b?

16. Approximate each number to the nearest hundredth.

a. $18\sqrt{2}$ **b.** $\dfrac{7\sqrt{3}}{3}$

17. Simplify each radical expression.

a. $\sqrt{81}$ **b.** $\sqrt{27}$

c. $\sqrt{3} \cdot \sqrt{3}$ **d.** $2 \cdot 5\sqrt{2}$

18. Rationalize the denominator.

a. $\dfrac{4}{\sqrt{3}}$ **b.** $\dfrac{16}{\sqrt{2}}$

NOTATION   *Complete the solution to solve the equation.*

19. Solve $8^2 + 6^2 = c^2$, where $c > 0$.

$$+\ 36 = c^2$$
$$= c$$
$$= c$$

20. Solve: $45 = a\sqrt{3}$

$$\frac{45}{} = \frac{a\sqrt{3}}{}$$

$$\frac{}{\sqrt{3}} = a$$

$$\frac{45}{\sqrt{3}} \cdot \frac{}{} = a$$

$$\frac{45\sqrt{3}}{} = a$$

$$\sqrt{3} = a$$

$$a \approx \qquad \text{Round to the nearest hundredth.}$$

21. Write the formula for the length c of the hypotenuse of an isosceles right triangle if a is the length of either one of the legs.

22. a. Write the formula for the length c of the hypotenuse of a 30°–60°–90° triangle if a is the length of the shorter leg.

　　b. Write the formula for the length b of the longer leg of a 30°–60°–90° triangle if a is the length of the shorter leg.

GUIDED PRACTICE

*Find the length of the hypotenuse of the right triangle shown below if it has the given side lengths. **See Examples 1 and 2.***

23. $a = 6$ ft and $b = 8$ ft

24. $a = 12$ mm and $b = 9$ mm

25. $a = 5$ m and $b = 12$ m

26. $a = 16$ in. and $b = 12$ in.

27. $a = 48$ mi and $b = 55$ mi

28. $a = 80$ ft and $b = 39$ ft

29. $a = 88$ cm and $b = 105$ cm

30. $a = 132$ mm and $b = 85$ mm

*Refer to the right triangle below. **See Example 3.***

31. Find b if $a = 10$ cm and $c = 26$ cm.

32. Find b if $a = 14$ in. and $c = 50$ in.

33. Find a if $b = 18$ m and $c = 82$ m.

34. Find a if $b = 9$ yd and $c = 41$ yd.

35. Find a if $b = 21$ m and $c = 29$ m.

36. Find a if $b = 16$ yd and $c = 34$ yd.

37. Find b if $a = 180$ m and $c = 181$ m.

38. Find b if $a = 630$ ft and $c = 650$ ft.

*The lengths of two sides of a right triangle are given. Find the missing side length. Approximate any exact answers that contain a radical to the nearest tenth. **See Example 4.***

39. $a = 5$ cm and $c = 7$ cm

40. $a = 4$ in. and $c = 8$ in.

41. $a = 12$ m and $b = 8$ m

42. $a = 10$ ft and $b = 4$ ft

43. $a = 9$ in. and $b = 3$ in.

44. $a = \sqrt{5}$ mi and $b = 7$ mi

45. $b = 4$ in. and $c = 6$ in.

46. $b = 9$ mm and $c = 12$ mm

*Find x. Then give the lengths of the other two sides of the right triangle. **See Example 5.***

47.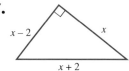

48.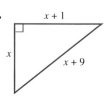

49. The longer leg of a right triangle is 1 unit longer than the shorter leg. If the hypotenuse is 2 units longer than the shorter leg, find the lengths of the sides of the triangle.

50. The shorter leg of a right triangle is 7 units shorter than the longer leg. If the hypotenuse is 1 unit longer than the longer leg, find the lengths of the sides of the triangle.

*Is a triangle with the following side lengths a right triangle? **See Example 6.***

51. 12, 14, 15

52. 15, 16, 22

53. 33, 56, 65

54. 20, 21, 29

*Find the length of the second leg of each right triangle. Then find the length of the hypotenuse. Give the exact answer and an approximation to the nearest hundredth. **See Example 7.***

55.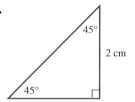

56.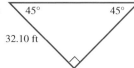

57. One leg of an isosceles right triangle is 3.2 feet long. Find the length of its hypotenuse. Give the exact answer and an approximation to the nearest hundredth.

58. One side of a square is $5\frac{1}{2}$ in. long. Find the length of its diagonal. Give the exact answer and an approximation to the nearest hundredth.

Find the length of a leg of each right triangle. Give the exact answer and an approximation to the nearest hundredth. **See Example 8.**

59.

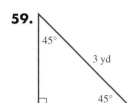

60.

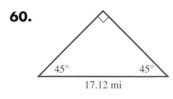

61. PHOTOGRAPHS   The diagonal of a square photograph measures 10 inches. Find the length of one of its sides. Give the exact answer and then an approximation to the nearest hundredth.

62. PARKING LOTS   The diagonal of a square parking lot is approximately 1,414 feet long. Find the length of one side of the parking lot. Round to the nearest foot.

Find the length of the hypotenuse and the longer leg of each triangle. Approximate any exact answers that contain a radical by rounding to the nearest hundredth. **See Example 9.**

63.

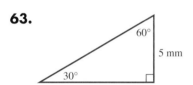

64.

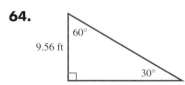

65. In a 30°–60°–90° triangle, the length of the leg opposite the 30° angle is 75 cm. Find the length of the leg opposite the 60° angle and the length of the hypotenuse.

66. In a 30°–60°–90° triangle, the length of the short leg is $5\sqrt{2}$ inches. Find the length of the hypotenuse and the length of the longer leg.

Find the missing lengths in each triangle. Approximate any exact answers that contain a radical by rounding to the nearest hundredth. **See Example 10.**

67.

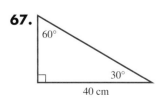

68.

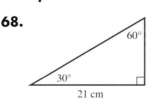

69. In a 30°–60°–90° triangle, the length of the leg opposite the 60° angle is 55 millimeters. Find the length of the leg opposite the 30° angle and the length of the hypotenuse.

70. In a 30°–60°–90° triangle, the length of the longer leg is 24 yards. Find the length of the hypotenuse and the length of the shorter leg.

Find the missing lengths in each triangle. Approximate any exact answers that contain a radical by rounding to the nearest hundredth. **See Example 11.**

71.

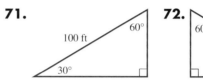

72.

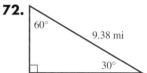

73. In a 30°–60°–90° triangle, the length of the hypotenuse is 1.5 feet. To the nearest hundredth, find the length of the shorter leg and the length of the longer leg.

74. In a 30°–60°–90° triangle, the length of the hypotenuse is $12\sqrt{3}$ inches. Find the length of the leg opposite the 30° angle and the length of the leg opposite the 60° angle.

TRY IT YOURSELF

The lengths of two sides of a right triangle shown below are given. Find the length of the missing side. Approximate any exact answers that contain a radical by rounding to the nearest tenth.

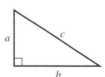

75. $a = \sqrt{15}$ in. and $b = 6$ in.

76. $a = 2\sqrt{2}$ yd. and $b = 9$ yd

77. $b = 3\sqrt{3}$ ft and $c = 5\sqrt{2}$ ft

78. $a = 6\sqrt{2}$ mi and $c = 10$ mi

79. Find the exact length of the diagonal (in blue) of one of the faces of the cube shown below.

80. Find the exact length of the diagonal (in green) of the cube shown below.

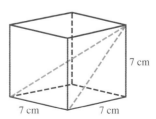

81. In a right triangle, one leg is 7 feet shorter than the other leg. The hypotenuse is 2 feet longer than the longer leg. Find the lengths of the sides of the triangle.

82. In a right triangle, one leg is 3 feet longer than the other leg. The hypotenuse is 3 feet longer than the longer leg. Find the lengths of the sides of the triangle.

APPLICATIONS

Give the exact answer. Approximate any exact answers that contain a radical by rounding to the nearest hundredth.

83. WASHINGTON, DC   The square in the illustration below shows the 100-square-mile site selected by George Washington in 1790 to serve as a permanent capital for the United States. In 1847, the part of the district lying on the west bank of the Potomac was returned to Virginia. Find the coordinates of each corner of the original square that outlined the District of Columbia.

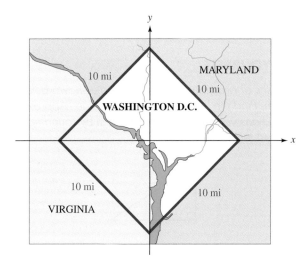

84. PAPER AIRPLANE   The figure below gives the directions for making a paper airplane from a square piece of paper with sides 8 inches long. Find the length *l* of the plane when it is completed.

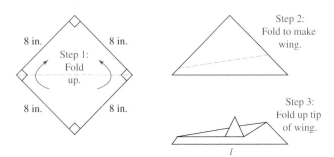

85. HARDWARE   The sides of the regular hexagonal nut shown below are 10 millimeters long. Find the height *h* of the nut.

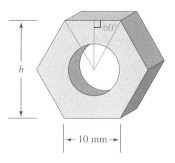

86. IRONING BOARD   Find the height *h* of the ironing board shown below.

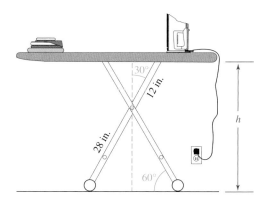

87. BASEBALL   The baseball diamond shown in the illustration below is a square, 90 feet on a side. If the third baseman fields a ground ball 10 feet directly behind third base, how far must he throw the ball to throw a runner out at first base?

88. BASEBALL   A shortstop fields a grounder at a point one-third of the way from second base to third base. (See the illustration below.) How far will he have to throw the ball to make an out at first base?

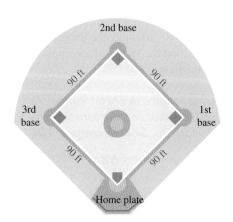

89. CLOTHESLINE   A pair of damp jeans are hung on a clothesline to dry, as shown below. They pull the center down 1 foot. By how much is the line stretched?

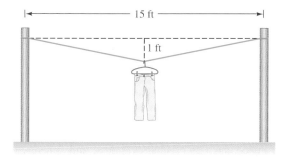

90. FIREFIGHTING   The base of the 37-foot ladder shown in the figure below is 9 feet from the wall. Will the top reach a window ledge that is 35 feet above the ground? Explain how you arrived at your answer.

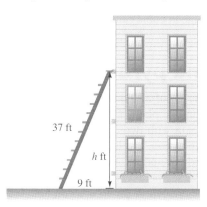

91. BOATING   The inclined ramp of the boat launch shown below is 8 meters longer than the "rise" of the ramp. The "run" is 7 meters longer than the "rise." How long are the three sides of the ramp?

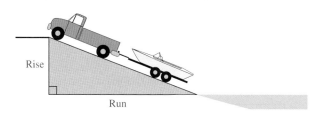

92. GARDENING TOOLS   The dimensions (in millimeters) of the triangular teeth of a pruning saw blade are shown below. Find each length.

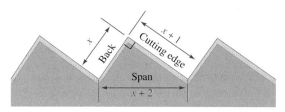

WRITING

93. State the Pythagorean theorem in your own words.

94. When the lengths of the sides of the triangle shown below are substituted into the equation $a^2 + b^2 = c^2$, the result is a false statement. Explain why.

$$a^2 + b^2 = c^2$$
$$2^2 + 4^2 = 5^2$$
$$4 + 16 = 25$$
$$20 = 25$$

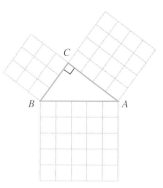

95. Explain why $\sqrt{3}a$ and $\sqrt{3a}$ are different.

96. In the figure below, equal-sized squares have been drawn on the sides of right triangle $\triangle ABC$. Explain how this figure demonstrates that $3^2 + 4^2 = 5^2$.

97. In the movie *The Wizard of Oz*, the scarecrow was in search of a brain. To prove that he had found one, he recited the following:

"The sum of the square roots of any two sides of an isosceles triangle is equal to the square root of the remaining side."

Unfortunately, this statement is not true. Correct it so that it states the Pythagorean theorem.

98. List the facts you learned about special right triangles in this section.

10 *Volume*

Objectives

1. Classify prisms and pyramids.
2. Identify cones and spheres.
3. Find the volume of prisms and pyramids.
4. Find the volume of cylinders, cones, and spheres.

We have studied ways to calculate the perimeter and the area of two-dimensional figures that lie in a plane, such as rectangles, triangles, and circles. Now we will consider three-dimensional figures that occupy space, such as prisms, cylinders, and spheres. In this section, we will introduce the vocabulary associated with these figures as well as the formulas that are used to find their volume. Volumes are measured in cubic units, such as cubic feet, cubic yards, or cubic centimeters. For example,

- We measure the capacity of a refrigerator in cubic feet.
- We buy gravel or topsoil by the cubic yard.
- We often measure amounts of medicine in cubic centimeters.

1. Classify prisms and pyramids.

In geometry, **space** is defined to be the set of all points. Space figures are geometric figures that contain points in more than one plane. One example of a space figure is the prism. To construct a prism, we begin with two congruent polygons lying in two parallel planes, as shown below. Each polygon, and its interior, is called a **base** of the prism. The sides of the bases are called base edges. When we join the corresponding vertices of each polygon with parallel line segments called **lateral edges,** parallelogram-shaped regions called **lateral faces** are created. A prism is a combination of its bases and its lateral faces. Its **height** h is the distance between the planes that contain its bases.

 CAUTION   There are always parts of a three-dimensional figure that cannot be seen from the position of the observer because they are covered by portions of the figure. The edges of these hidden parts are indicated by dashed lines in a drawing.

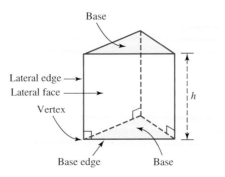

Prisms are classified according to the shape of their bases. Because the prism in the previous figure has a three-sided base, it is called a *triangular* prism. In addition, if the lateral edges of a prism are perpendicular to its bases, it is called a **right prism.** Therefore, that prism is more specifically a *right* triangular prism. Prisms that are not right prisms are called **oblique.** Several other examples of prisms are shown below.

Right square prism   Oblique rectangular prism   Right pentagonal prism

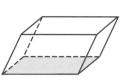

 CAUTION A right rectangular prism (a prism that has two bases that are rectangles and four lateral faces that are also rectangles) is commonly referred to as a **rectangular solid.** This name is misleading because a prism is not solid all the way through like a brick. Visualize a prism as an empty shoe box. If the bases and lateral faces of a prism are squares, it is called a **cube.** Examples of these two special types of prisms are shown below.

Rectangular solid   Cube

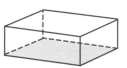

Another type of space figure is the **circular cylinder.** Circular cylinders are constructed in a manner similar to prisms. However, the bases of cylinders are congruent circles (circles with the same area). If the segment joining the centers of the circular bases is perpendicular to the bases, the figure is a **right circular cylinder.** Cylinders that are not right cylinders are referred to as **oblique.** The **height** h of a circular cylinder is the distance between the planes that contain the bases. Two examples of circular cylinders are shown below.

Right circular cylinder   Oblique circular cylinder

To construct another type of space figure, called a **pyramid,** we begin with a polygonal region (the **base**) and a point not in the plane of the region (the **vertex**), as shown below. Line segments join the vertex of the pyramid to the vertices of the base. These segments are called **lateral edges.** Each triangular region determined by the edge of the base and two lateral edges is called a **lateral face.** A pyramid is a combination of its base and its lateral faces. Its **height** h is the perpendicular distance from the vertex to the plane of the base.

Right regular hexagonal pyramid

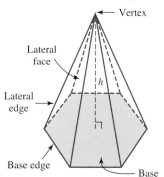

Pyramids are classified in the same way as prisms—by the shape of the base, and as either right or oblique. The vertex of a **right pyramid** is directly above the center of the

base. A pyramid whose base is a regular polygon and whose vertex is equidistant from each vertex of the base is called a **regular pyramid.** Therefore, the pyramid shown in the figure on the previous page, with a 6-sided base, is a *right regular* hexagonal pyramid. Two other examples of pyramids are shown below.

Right square pyramid

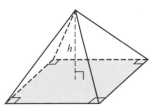

Oblique triangular pyramid

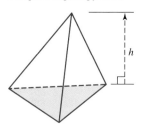

EXAMPLE 1 Refer to the prism below.

a. How many bases does the figure have? What shape are they?

b. How many lateral faces does it have? What shape are they?

c. What is the specific name of the figure?

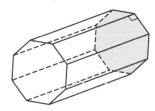

Strategy
We will look for two congruent polygons that lie in parallel planes.

Why
Two such polygons serve as the bases of the prism.

Solution

a. This figure has two bases—one directly facing the observer and one at the back that is shaded and partially outlined with dashed-line segments. Each base has 8 sides; they are octagons.

b. The figure has 8 lateral faces. Each is a rectangle.

c. Because the lateral edges are perpendicular to the bases and because the bases have 8 sides, this is a right octagonal prism.

Self Check 1
Refer to the pyramid below.

a. How many bases does the figure have? What shape is the base?

b. How many lateral faces does the figure have? What shape are they?

c. What is the specific name of the figure?

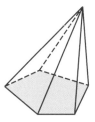

Answers: **a.** 1, pentagon **b.** 5, triangles **c.** oblique pentagonal pyramid

Now Try
Problems 29 and 33

2. Identify cones and spheres.

To construct a **circular cone,** we begin with a circle in one plane and then choose a point, called the **vertex,** that is not in the plane. The circular region, together with the set of all segments connecting the vertex to a point on the circle, forms the cone. The variable r is normally used to represent the **radius** of the base, and the **height** h is the perpendicular distance from the vertex to the plane of the base. If a segment from the vertex of the cone to the center of the base is perpendicular to the base, the cone is called a **right circular cone.** Otherwise, it is said to be **oblique.** Two examples of circular cones are shown below.

Right circular cone

Oblique circular cone

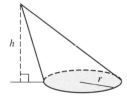

The final type of space figure that we will discuss is the **sphere.** A sphere is the set of all points in space that are a given distance, called the **radius,** from a given point, called the **center.** The radius of a sphere is usually represented by the variable *r*, as shown below.

A **hemisphere** is an exact half of a sphere.

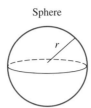

Sphere

Hemisphere

3. Find the volume of prisms and pyramids.

The **volume** of a three-dimensional figure is a measure of its capacity. The following illustration shows two common units of volume: cubic inches (in.^{3}) and cubic centimeters (cm^3).

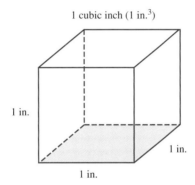

1 cubic inch (1 in.^{3})

1 in.

1 in.

1 in.

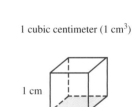

1 cubic centimeter (1 cm^3)

1 cm

1 cm

1 cm

The volume of a figure can be thought of as the number of cubic units that will fit within its boundaries. If we divide the right rectangular prism (shown in black) in the figure below into cubes, each cube represents a volume of 1 cm^3. Because there are 2 levels with 12 cubes on each level, the volume of the prism is 24 cm^3.

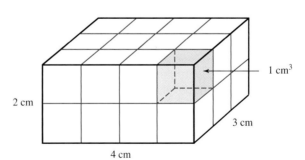

2 cm

3 cm

4 cm

1 cm^3

EXAMPLE 2 How many cubic inches are there in 1 cubic foot?

Strategy
A figure is helpful to solve this problem. We will draw a cube and divide each of its sides into 12 equally long parts.

Why
Since a cubic foot is a cube with each side measuring 1 foot, each side also measures 12 inches.

Self Check 2
How many cubic centimeters are in 1 cubic meter?

Solution

The figure on the right helps us understand the situation. Note that each level of the cubic foot contains 12 · 12 cubic inches and that the cubic foot has 12 levels. We can use multiplication to count the number of cubic inches contained in the figure. There are

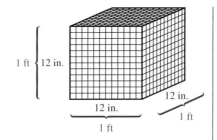

$$12 \cdot 12 \cdot 12 = 1,728$$

cubic inches in 1 cubic foot.

Answer: $1,000,000 \text{ cm}^3$

Now Try
Problem 37

In practice, we do not find volumes by counting cubes. Instead, we use the formulas shown in the following table. Note that several of the volume formulas involve the variable B. It represents the area of the base of the figure.

Cube	**Rectangular solid**	**Sphere**
	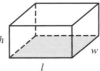	
$V = s^3$	$V = lwh$	$V = \frac{4}{3}\pi r^3$
where s is the length of a side	where l is the length, w is the width, and h is the height	where r is the radius

Prism	**Pyramid**
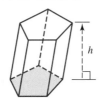	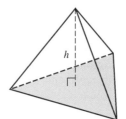
$V = Bh$	$V = \frac{1}{3}Bh$
where B is the area of the base and h is the height	where B is the area of the base and h is the height

Circular cylinder	**Circular cone**
 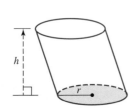	
$V = Bh$ or $V = \pi r^2 h$	$V = \frac{1}{3}Bh$ or $V = \frac{1}{3}\pi r^2 h$
where B is the area of the base, h is the height, and r is the radius	where B is the area of the base, h is the height, and r is the radius

CAUTION   The height of a geometric solid is always measured along a line perpendicular to its base.

EXAMPLE 3   *Storage tanks.*   An oil storage tank is in the form of a rectangular solid with dimensions 17 feet by 10 feet by 8 feet. (See the figure below.) Find its volume.

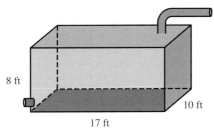

8 ft

10 ft

17 ft

Strategy
We will substitute 17 for *l*, 10 for *w*, and 8 for *h* in the formula $V = lwh$ and evaluate the right side.

Why
The variable *V* represents the volume of a rectangular solid.

Solution

$V = lwh$	This is the formula for the volume of a rectangular solid.
$V = 17(10)(8)$	Substitute 17 for *l*, the length, 10 for *w*, the width, and 8 for *h*, the height of the tank.
$= 1,360$	Do the multiplication.

The volume of the tank is 1,360 ft³.

Self Check 3
Find the volume of a rectangular solid with dimensions 8 meters by 12 meters by 20 meters.

Answer: 1,920 m³

Now Try
Problem 41

EXAMPLE 4   Find the volume of the right triangular prism shown on the right.

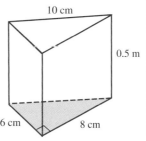

10 cm

0.5 m

6 cm   8 cm

Strategy
We will convert the height of the triangular prism to centimeters and then use the formula $V = Bh$ to find its volume.

Why
The dimensions of the figure must be expressed in the same units.

Solution
Since there are 100 centimeters in 1 meter, the height of the triangular prism in centimeters is

$0.5 \text{ m} = 0.5 \text{ m} \cdot \dfrac{100 \text{ cm}}{1 \text{ m}}$	Multiply by 1: $\frac{100 \text{ cm}}{1 \text{ m}} = 1$.
$= 0.5(100 \text{ cm})$	Remove the common units of meters in the numerator and denominator.
$= 50 \text{ cm}$	

The area of the triangular base of the triangular prism is $\frac{1}{2}(6)(8) = 24$ square centimeters. To find its volume, we proceed as follows:

$V = Bh$	This is the formula for the volume of a triangular prism.
$V = 24(50)$	Substitute 24 for *B* and 50 for *h*.
$= 1,200$	Do the multiplication.

The volume of the triangular prism is 1,200 cm³.

Self Check 4
Find the volume of the right triangular prism shown below.

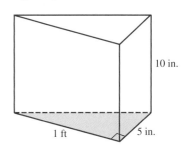

10 in.

1 ft   5 in.

Answer: 300 in.³

Now Try
Problem 45

CAUTION In Example 4, note that the 10 cm measurement was not used in the calculation of the volume.

EXAMPLE 5 Find the volume of the pyramid shown below.

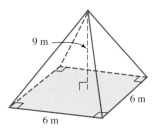

Strategy
First, we will find the area of the square base of the pyramid.

Why
The volume of a pyramid is $\frac{1}{3}$ of the product of the area of its base and its height.

Solution
Since the base is a square with each side 6 meters long, the area of the base is $(6 \text{ m})^2$, or 36 m^2. To find the volume of the pyramid, we proceed as follows:

$$V = \frac{1}{3}Bh \qquad \text{This is the formula for the volume of a pyramid.}$$

$$V = \frac{1}{3}(36)(9) \qquad \text{Substitute 36 for } B \text{, the area of the base, and 9 for } h \text{, the height}$$

$$= 12(9) \qquad \text{Multiply: } \frac{1}{3}(36) = \frac{36}{3} = 12.$$

$$= 108 \qquad \text{Complete the multiplication.}$$

The volume of the pyramid is 108 m^3.

Self Check 5
Find the volume of the pyramid shown below.

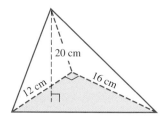

Answer: 640 cm^3

Now Try
Problem 49

EXAMPLE 6 *Egyptian history.* The largest of all pyramids built by the Egyptians was the Great Pyramid. Constructed about 4,000 years ago near Cairo, it is the only one of the Seven Wonders of the Ancient World still in existence. Its square base covers an area of $571,536 \text{ ft}^2$, and its volume is an astounding $91,636,272 \text{ ft}^3$. Find its height.

Strategy
We will substitute 91,636,272 for V and 571,536 for B in the formula $V = \frac{1}{3}Bh$.

Why
We can then solve the equation for h, the height of the pyramid.

Solution

$$V = \frac{1}{3}Bh \qquad \text{This is the formula for the volume of a pyramid.}$$

$$91,636,272 = \frac{1}{3}(571,536)h \qquad \text{Substitute 91,636,272 for } V \text{, the volume, and } 571,536 \text{ for } B \text{, the area of the base.}$$

$$3 \cdot 91,636,272 = 3 \cdot \frac{1}{3}(571,536)h \qquad \text{To clear the equation of the fraction, multiply both sides by 3.}$$

$$274,908,816 = 571,536h \qquad \text{Multiply: } 3 \cdot \frac{1}{3} = 1.$$

$$\frac{274,908,816}{571,536} = \frac{571,536h}{571,536} \qquad \text{To isolate } h \text{, undo the multiplication by 571,536 by dividing both sides by 571,536.}$$

$$481 = h \qquad \text{Use a calculator to do the division.}$$

The height of the Great Pyramid is 481 ft.

Now Try
Problem 57

4. Find the volume of cylinders, cones, and spheres.

EXAMPLE 7 Find the volume of the circular cylinder shown on the right. Give the exact answer and an approximation to the nearest hundredth.

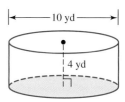

Strategy

First, we will find the radius of the circular base of the cylinder.

Why

To use the formula for the volume of a cylinder, $V = \pi r^2 h$, we need to know r, the radius of the base.

Solution

Since a radius is one-half of the diameter of the circular base, $r = \frac{1}{2} \cdot 6$ cm $= 3$ cm. From the figure, we see that the height of the cylinder is 10 cm. To find the volume of the cylinder, we proceed as follows.

$V = \pi r^2 h$	This is the formula for the volume of a cylinder.
$V = \pi(3)^2(10)$	Substitute 3 for r, the radius of the base, and 10 for h, the height.
$V = \pi(9)(10)$	Evaluate the exponential expression: $(3)^2 = 9$.
$= 90\pi$	Multiply: $(9)(10) = 90$.
≈ 282.7433388	Use a calculator to do the multiplication.

The exact volume of the cylinder is 90π cm^3. To the nearest hundredth, the volume is 282.74 cm^3.

Self Check 7

Find the volume of the circular cylinder shown below. Give the exact answer and an approximation to the nearest hundredth.

Answer: 100π yd$^3 \approx 314.16$ yd^3

Now Try

Problem 61

EXAMPLE 8 Find the volume of the circular cone shown on the right. Give the exact answer and an approximation to the nearest hundredth.

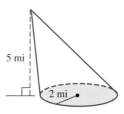

Strategy

We will substitute 4 for r and 6 for h in the formula $V = \frac{1}{3}\pi r^2 h$ and evaluate the right side.

Why

The variable V represents the volume of a cone.

Solution

$V = \frac{1}{3}\pi r^2 h$	This is the formula for the volume of a cone.
$V = \frac{1}{3}\pi(4)^2(6)$	Substitute 4 for r, the radius of the base, and 6 for h, the height.
$= \frac{1}{3}\pi(16)(6)$	Evaluate the exponential expression: $(4)^2 = 16$.
$= 2\pi(16)$	Multiply: $\frac{1}{3}(6) = 2$.
$= 32\pi$	Multiply: $2(16) = 32$.
≈ 100.5309649	Use a calculator to do the multiplication.

The exact volume of the cone is 32π ft^3. To the nearest hundredth, the volume is 100.53 ft^3.

Self Check 8

Find the volume of the circular cone shown below. Give the exact answer and an approximation to the nearest hundredth.

Answer: $\frac{20}{3}\pi$ mi$^2 \approx 20.94$ mi^2

Now Try

Problem 65

EXAMPLE 9 *Water towers.* How many cubic feet of water are needed to fill the spherical water tank shown on the right? Give the exact answer and an approximation to the nearest tenth.

15 ft

Strategy
We will substitute 15 for r in the formula $V = \frac{4}{3}\pi r^3$ and evaluate the right side.

Why
The variable V represents the volume of a sphere.

Solution

$$V = \frac{4}{3}\pi r^3 \qquad \text{This is the formula for the volume of a sphere.}$$

$$V = \frac{4}{3}\pi(\mathbf{15})^3 \qquad \text{Substitute 15 for } r, \text{ the radius of the sphere.}$$

$$= \frac{4}{3}\pi(3{,}375) \qquad \text{Evaluate the exponential expression: } (15)^3 = 3{,}375.$$

$$= \frac{13{,}500}{3}\pi \qquad \text{Multiply: } 4(3{,}375) = 13{,}500.$$

$$= 4{,}500\pi \qquad \text{Divide: } \frac{13{,}500}{3} = 4{,}500.$$

$$V \approx 14{,}137.16694 \qquad \text{Use a calculator to do the multiplication.}$$

The tank holds exactly $4{,}500\pi$ ft^3 of water. To the nearest tenth, this is $14{,}137.2$ ft^3.

Self Check 9
Find the volume of a spherical water tank with radius 7 meters. Give the exact answer and an approximation to the nearest tenth.

Answer: $\frac{1{,}372}{3}\pi$ m$^3 \approx 1{,}436.8$ m^3

Now Try
Problem 69

EXAMPLE 10 A sphere has a volume of 36π in.^{3}. What is its diameter?

Strategy
We will substitute 36π for V in the formula $V = \frac{4}{3}\pi r^3$ and solve for r.

Why
We can then double the value of r to find the diameter of the sphere.

Solution

$$V = \frac{4}{3}\pi r^3 \qquad \text{This is the formula for the volume of a sphere.}$$

$$36\pi = \frac{4}{3}\pi r^3 \qquad \text{Substitute } 36\pi \text{ for } V, \text{ the volume.}$$

$$3 \cdot 36\pi = 3 \cdot \frac{4}{3}\pi r^3 \qquad \text{To clear the equation of the fraction, multiply both sides by 3.}$$

$$108\pi = 4\pi r^3 \qquad \text{Do each multiplication.}$$

$$\frac{108\pi}{4\pi} = \frac{4\pi r^3}{4\pi} \qquad \text{To isolate } r^3, \text{ undo the multiplication by } 4\pi \text{ by dividing both sides by } 4\pi.$$

$$\frac{\overset{1}{\cancel{4}} \cdot 27 \cdot \overset{1}{\cancel{\pi}}}{\underset{1}{\cancel{4}} \cdot \underset{1}{\cancel{\pi}}} = r^3 \qquad \text{Simplify each fraction.}$$

$$27 = r^3$$

$$r^3 = 27 \qquad \text{Reverse the sides of the equation so that } r^3 \text{ is on the left.}$$

$$r = 3 \qquad \text{Ask: "What number cubed is 27?" The answer is 3.}$$

Since the radius of the sphere is 3 inches, its diameter is $2 \cdot 3$ inches $= 6$ inches.

Self Check 10
A sphere has a volume of 288π ft^3. What is its diameter?

Answer: 12 ft

Now Try
Problem 73

Using Your Calculator: **Volume of a silo**

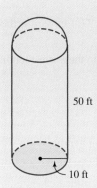

A silo is a structure used for storing grain. The silo shown on the left is a cylinder 50 feet tall topped with a dome in the shape of a hemisphere. To find the volume of the silo, we add the volume of the cylinder to the volume of the dome.

$$\text{Volume}_{\text{cylinder}} + \text{Volume}_{\text{dome}} = (\text{Area}_{\text{cylinder's base}})(\text{Height}_{\text{cylinder}}) + \frac{1}{2}(\text{Volume}_{\text{sphere}})$$

50 ft

$$= \pi r^2 h + \frac{1}{2}\left(\frac{4}{3}\pi r^3\right)$$

$$= \pi r^2 h + \frac{2\pi r^3}{3} \qquad \text{Multiply and simplify: } \frac{1}{2}\left(\frac{4}{3}\pi r^3\right) = \frac{4}{6}\pi r^3 = \frac{2\pi r^3}{3}.$$

10 ft

$$= \pi(10)^2\,(50) + \frac{2\pi(10)^3}{3} \qquad \text{Substitute 10 for } r \text{ and 50 for } h.$$

We can use a scientific calculator to make this calculation.

Keystrokes $\boxed{\pi}\ \boxed{\times}\ 10\ \boxed{x^2}\ \boxed{\times}\ 50\ \boxed{+}\ \boxed{(}\ 2\ \boxed{\times}\ \boxed{\pi}\ \boxed{\times}\ 10\ \boxed{y^x}\ 3\ \boxed{)}\ \boxed{\div}\ 3\ \boxed{=}$

$$\boxed{17802.35837}$$

The volume of the silo is approximately 17,802 ft^3.

Now Try
Problem 103

STUDY SET Section 10

VOCABULARY *Fill in the blanks.*

1. In geometry, _____ is defined to be the set of all points.

2. Space figures are geometric figures that contain points in more than one _____.

3. If the lateral edges of a prism are perpendicular to its bases, it is called a _____ prism. Prisms that are not right prisms are called _____.

4. A right rectangular prism is commonly referred to as a _____ solid.

5. If the bases and lateral faces of a prism are squares, the prism is called a _____.

6. A _____ is the set of all points in space that are a given distance, called the radius, from a given point, called the center. A _____ is an exact half of a sphere.

7. The height of a pyramid or cone is the _____ distance from the vertex to the plane of the base.

8. The _____ of a three-dimensional figure is a measure of its capacity.

9. The volume of a figure can be thought of as the number of _____ units that will fit within its boundaries.

10. Copy the figure on the right. Label it completely using the following words:

base   base edge   lateral edge
height   vertex   lateral face

CONCEPTS

11. Give the complete name of each figure.

a.

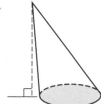

b.

c.

d.

e.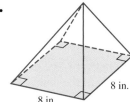

8 in.
8 in.

f.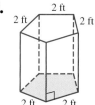

2 ft
2 ft   2 ft
2 ft   2 ft

12. Draw a right hexagonal prism. Label the height *h*, the bases, and a lateral face.

13. Draw a cube. Label each base and a lateral face.

14. Draw a right circular cylinder. Label the height *h* and a radius *r*.

15. Draw an oblique square pyramid. Label the height *h*, the vertex, and the base.

16. Draw a right regular pentagonal pyramid. Label the height *h*, the vertex, and the base.

17. Draw an oblique circular cone. Label the height *h*, the vertex, and radius *r*.

18. Draw a sphere. Label a radius *r*.

19. Which of the following are acceptable units with which to measure volume?

ft^2	mi^3	seconds	days
cubic inches	mm	square yards	in.
pounds	cm^2	meters	m^3

20. In the figure on the right, the unit of measurement of length used to draw the figure is the inch.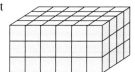

   a. What is the area of the base of the figure?

   b. What is the volume of the figure?

21. Which geometric concept (perimeter, circumference, area, or volume) should be applied when measuring each of the following?

   a. The distance around a checkerboard

   b. The size of a trunk of a car

   c. The amount of paper used for a postage stamp

   d. The amount of storage in a cedar chest

   e. The amount of beach available for sunbathing

   f. The distance the tip of a propeller travels

22. Complete the table.

Figure	Volume Formula
Cube	
Rectangular solid	
Prism	
Circular cylinder	
Pyramid	
Circular cone	
Sphere	

23. Evaluate each expression. Leave π in the answer.

   a. $\frac{1}{3}\pi(25)6$ **b.** $\frac{4}{3}\pi(125)$

24. a. Evaluate $\frac{1}{3}\pi r^2 h$ for $r = 2$ and $h = 27$. Leave π in the answer.

   b. Approximate your answer to part a to the nearest tenth.

NOTATION

25. a. What does "in.^{3}" mean?

   b. Write "one cubic centimeter" using symbols.

26. In the formula $V = \frac{1}{3}Bh$, what does *B* represent?

27. In a drawing, what does the symbol $\llcorner$ indicate?

28. Redraw the figure below using dashed lines to show the hidden edges.

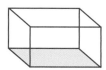

GUIDED PRACTICE   *Answer the following questions for each figure shown below.* **See Example 1.**

   a. How many bases does the figure have? What shape are they?

   b. How many lateral faces does it have? What shape are they?

   c. What is the specific name of the figure?

29. 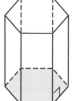   **30.**

31. 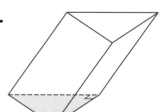   **32.**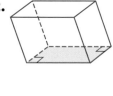

Answer the following questions for each figure shown in Problems 33–36. **See Example 1.**

   a. How many bases does the figure have? What shape is the base?

   b. How many lateral faces does the figure have? What shape are they?

   c. What is the specific name of the figure?

33.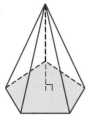

34.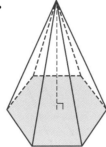

35.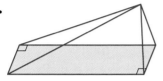

36.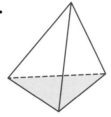

Convert from one unit of measurement to another. **See Example 2.**

37. How many cubic feet are in 1 cubic yard?

38. How many cubic decimeters are in 1 cubic meter?

39. How many cubic meters are in 1 cubic kilometer?

40. How many cubic inches are in 1 cubic yard?

Find the volume of each figure. **See Example 3.**

41.

42.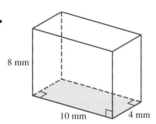

43.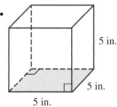

44.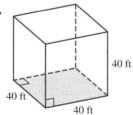

Find the volume of each figure. **See Example 4.**

45.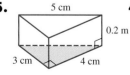

46.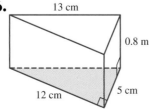

47.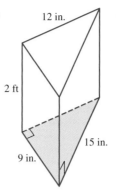

48.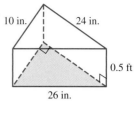

Find the volume of each figure. **See Example 5.**

49.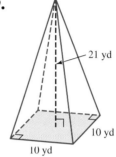

50.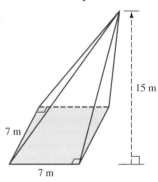

51.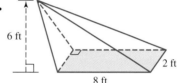

52.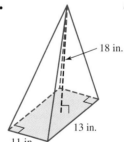

53.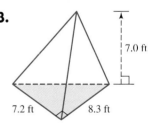

54.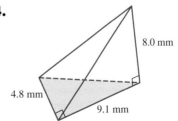

55.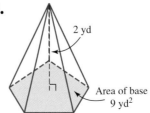

56.

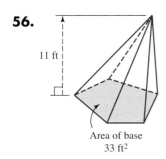

Area of base
33 ft²

Find the height of each square pyramid with the given dimensions. **See Example 6.**

57. Area of the base = 361 in.²; volume = 3,249 in.³

58. Area of the base = 841 m²; volume = 4,205 m³

59. Area of the base = 16,384 cm²; volume = 311,296 cm³

60. Area of the base = 63,504 in.²; volume = 1,375,920 in.³

Find the volume of each cylinder. Give the exact answer and an approximation to the nearest hundredth. **See Example 7.**

61.

62.

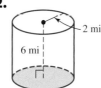

63.

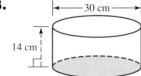

64.

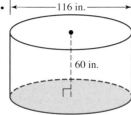

Find the volume of each cone. Give the exact answer and an approximation to the nearest hundredth. **See Example 8.**

65.

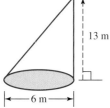

66.

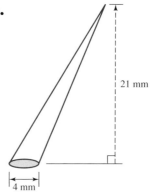

67.

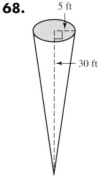

68.

Find the volume of each sphere. Give the exact answer and an approximation to the nearest tenth. **See Example 9.**

69.

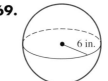

70.

71.

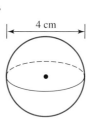

72.

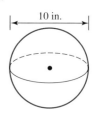

What is the diameter of a sphere with the given volume? **See Example 10.**

73. $\frac{32}{3}\pi$ cm³

74. $\frac{256}{3}\pi$ mm³

75. 972π ft³

76. $2,304\pi$ m³

TRY IT YOURSELF   *Find the volume of each figure. If an exact answer contains π, approximate to the nearest hundredth.*

77. A hemisphere with a radius of 9 inches

78. A hemisphere with a diameter of 22 feet

79. A cylinder with a height of 12 meters and a circular base with a radius of 6 meters

80. A cylinder with a height of 4 meters and a circular base with a diameter of 18 meters

81. A rectangular solid with dimensions of 3 cm by 4 cm by 5 cm

82. A rectangular solid with dimensions of 5 m by 8 m by 10 m

83. A cone with a height of 12 centimeters and a circular base with a diameter of 10 centimeters

84. A cone with a height of 3 inches and a circular base with a radius of 4 inches

85. A pyramid with a square base 10 meters on each side and a height of 12 meters

86. A pyramid with a square base 6 inches on each side and a height of 4 inches

87. A prism whose base is a right triangle with legs 3 meters and 4 meters long and whose height is 8 meters

88. A prism whose base is a right triangle with legs 5 feet and 12 feet long and whose height is 25 feet

89. The volume of a cube is 27 ft^3. What is the length of a side of the cube?

90. The area of the base of a prism is 16 in.^{2}, and its volume is 88 in.^{3}. What is the height of the prism?

91. The volume of a circular cylinder is 200π yd^3, and the radius of its circular base is 5 yd. What is its height?

92. The volume of a circular cylinder is 144π m^3, and its height is 9 m. What is the radius of its circular base?

93. The volume of a pyramid is 95 ft^3, and the area of its base is 19 ft^2. What is its height?

94. The volume of a sphere is $\frac{500}{3}\pi$ mi^3. Find its diameter.

Find the volume of each figure. Express your answer in ft^3.

95.
60 in.
2 ft
1 yd

96.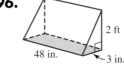
2 ft
48 in.
3 in.

Find the volume of each figure. Express your answer in m^3. Give the exact answer and an approximation to the nearest hundredth.

97.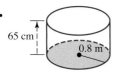
65 cm
0.8 m

98.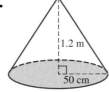
1.2 m
50 cm

Find the volume of each figure. Give the exact answer and an approximation to the nearest hundredth.

99. Right regular hexagonal prism

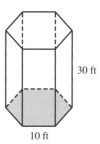
30 ft
10 ft

100. Right regular triangular pyramid

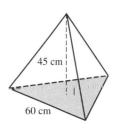
45 cm
60 cm

101.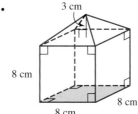
3 cm
8 cm
8 cm
8 cm

102.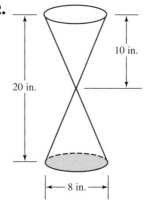
10 in.
20 in.
8 in.

103.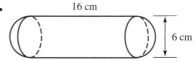
16 cm
6 cm

104.

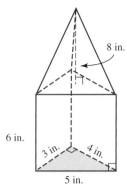

APPLICATIONS *Solve each problem. If an exact answer contains π, approximate the answer to the nearest hundredth.*

105. SWEETENERS   A sugar cube is $\frac{1}{2}$ inch on each edge. How much volume does it occupy?

106. VENTILATION   A classroom is 40 feet long, 30 feet wide, and 9 feet high. Find the number of cubic feet of air in the room.

107. WATER HEATERS   Complete the advertisement for the high-efficiency water heater shown below.

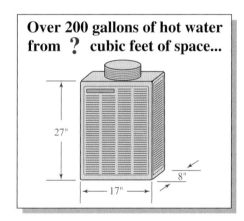

Over 200 gallons of hot water from ? cubic feet of space...

108. REFRIGERATORS   The largest refrigerator advertised in a J. C. Penny catalog has a capacity of 25.2 cubic feet. How many cubic inches is this?

109. TANKS   A cylindrical oil tank has a diameter of 6 feet and a length of 7 feet. Find the volume of the tank.

110. DESSERTS   A restaurant serves pudding in a conical dish that has a diameter of 3 inches. If the dish is 4 inches deep, how many cubic inches of pudding are in each dish?

111. HOT-AIR BALLOONS   The lifting power of a spherical balloon depends on its volume. How many cubic feet of gas will a balloon hold if it is 40 feet in diameter?

112. CEREAL BOXES   A box of cereal measures 3 inches by 8 inches by 10 inches. The manufacturer plans to market a smaller box that measures $2\frac{1}{2}$ by 7 by 8 inches. By how much will the volume be reduced?

113. ENGINES   The *compression ratio* of an engine is the volume in one cylinder with the piston at bottom-dead-center (B.D.C.), divided by the volume with the piston at top-dead-center (T.D.C.). From the data given in the following figure, what is the compression ratio of the engine? Use a colon to express your answer.

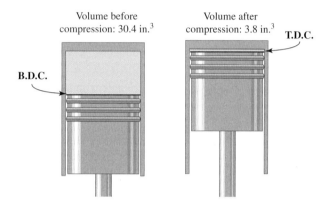

Volume before compression: 30.4 in.^{3}

Volume after compression: 3.8 in.^{3}

114. GEOGRAPHY   Earth is not a perfect sphere but is slightly pear-shaped. To estimate its volume, we will assume that it is spherical, with a diameter of about 7,926 miles. What is its volume, to the nearest one billion cubic miles?

115. BIRDBATHS
  a. The bowl of the birdbath shown below is in the shape of a hemisphere. Find its volume.
  b. If 1 gallon of water occupies 231 cubic inches of space, how many gallons of water does the birdbath hold? Round to the nearest tenth.

30 in.

116. CONCRETE BLOCKS   Find the number of cubic inches of concrete used to make the hollow block shown below.

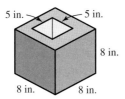

WRITING

117. What is meant by the *volume* of a cube?

118. The stack of 3 × 5 index cards shown in figure (a) forms a right rectangular prism, with a certain volume. If the stack is pushed to lean to the right, as in figure (b), a new prism is formed. How will its volume compare to the volume of the right rectangular prism? Explain your answer.

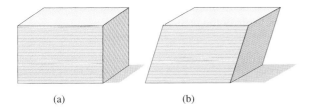

(a)   (b)

119. Are the units used to measure area different from the units used to measure volume? Explain.

120. The dimensions (length, width, and height) of one rectangular solid are entirely different numbers from the dimensions of another rectangular solid. Would it be possible for the rectangular solids to have the same volume? Explain.

121. If the height of a pyramid is doubled, how does its volume change? Explain your reasoning.

122. If the radius of a cylinder is doubled, how does its volume change? Explain your reasoning.

11 Surface Area

Objectives

1. Find the total surface area of prisms.
2. Find the total surface area of cylinders.
3. Find the total surface area of pyramids.
4. Find the total surface area of cones.
5. Find the surface area of spheres.

We have previously used formulas to calculate the areas of two-dimensional figures that lie in a plane, such as squares, rectangles, and circles. Now we will extend this concept to three-dimensional figures, such as prisms, cylinders, and spheres, to find their *surface area*. The ability to compute surface area is necessary when determining the amount of material that is needed to make a cardboard box, an aluminum can, or a plastic beach ball.

1. Find the total surface area of prisms.

The cardboard box shown in figure (a) is in the shape of a right rectangular prism (rectangular solid). Recall that the top and bottom are called bases, and the front, back, left end, and right end are called lateral faces of the prism.

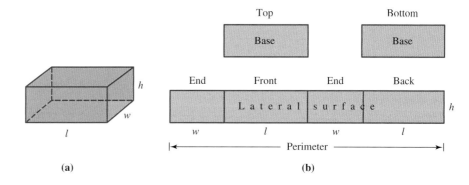

(a)   **(b)**

The **total surface area** *TSA* of the box is the sum of the area of its bases *and* the areas of its four lateral faces. It gives us a measure of the amount of cardboard needed to make it. To derive a formula for the total surface area of such a figure, we disassemble the box and lay the pieces of cardboard out flat, as shown in figure (b). We observe the following.

- The bases of the box are congruent. If B is the area of one base, then $B + B = 2B$ is the sum of the areas of both of its bases.

- The sum of the areas of the four lateral faces of the box is called its **lateral surface area** *LSA*. One way to find the *LSA* is to compute the area of each of the lateral faces and then add them. Or we can use the fact that, when laid out flat, the lateral faces form one large rectangle. (See figure (b).) The width of the rectangle is the height of the box; the length of the rectangle is the perimeter of the base of the box. To find the lateral surface area, we can simply multiply the height h of the box by the perimeter p of its base: $LSA = hp$.

These observations suggest the following formula.

Total surface area of a right prism

The total surface area *TSA* of a right prism is the sum of the area of its bases and its lateral surface area. If h is the height, p the perimeter of each base, and B the area of each base, then the total surface area of a right prism is given by

$$TSA = 2B + hp$$

CAUTION   As with any type of measurement of area, surface area is measured in square units, such as square feet (ft^2), square inches (in.^{2}), and square centimeters (cm^2).

EXAMPLE 1   Find the total surface area of the right prism shown on the right.

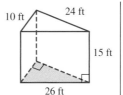

Strategy
We will find the area of one of the bases of the prism, the perimeter of a base, and the height of the prism.

Why
The total surface area of a prism is the sum of the area of its bases and its lateral surface area. To calculate the lateral surface area, we need to know the perimeter of a base and the height of the prism.

Solution
To find the area of one of the triangular bases of the prism, we substitute 10 for b and 24 for h in the formula for the area of a triangle.

$$A = \frac{1}{2}bh$$

$A = \frac{1}{2}(10)(24)$ One leg of the right triangle is the base, and the other leg is the height of the triangle.

$\quad = 5(24)$ Multiply: $\frac{1}{2}(10) = 5$.

$\quad = 120$ Complete the multiplication.

The area of one base of the prism is 120 ft^2.
   The perimeter p (in feet) of a base is

$$p = 10 + 24 + 26 = 60$$

To find the total surface area of the prism, we proceed as follows.

$TSA = 2B + hp$ This is the formula for the surface area of a prism.

$TSA = 2(120) + 15(60)$ Substitute 120 for B, 15 for h, and 60 for p.

$TSA = 240 + 900$ The area of the bases is 240 ft^2, and the *LSA* is 900 ft^2.

$TSA = 1,140$ Do the addition.

The total surface area of the prism is 1,140 ft^2.

EXAMPLE 2   Find the area of one base of the right regular octagonal prism shown on the right if its total surface area is 480 m^2.

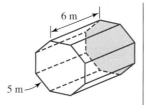

Strategy
We will substitute values for *TSA*, h, and p in the formula $TSA = 2B + hp$.

Why
We can then solve the equation for B, the area of one base of the prism.

Solution
The prism is lying on one of its lateral faces. Since the base is a regular octagon, each of the eight sides is 5 m long, and the perimeter of the base is $8 \cdot 5 = 40$ m. We substitute

---

Self Check 1
Find the total surface area of the cube shown below.

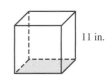

Answer: 726 in.^{2}

Now Try
Problem 23   ■

Self Check 2
Find the area of one base of the right regular octagonal prism shown below if its total surface area is 780 ft^2.

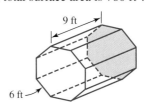

480 for *TSA*, 6 for *h*, and 40 for *p* in the formula for the total surface area of a prism and solve for *B*, the area of one base.

$$TSA = 2B + hp$$
$$480 = 2B + 6(40)$$
$$480 = 2B + 240 \quad \text{Do the multiplication.}$$
$$240 = 2B \qquad \text{To isolate the variable term } 2B, \text{ subtract } 240 \text{ from both sides.}$$
$$120 = B \qquad \text{To isolate the variable } B, \text{ undo the multiplication by 2 by dividing both sides by 2.}$$

The area of one base of the prism is 120 m^2.

Answer: 174 ft^2

Now Try
Problem 31

2. Find the total surface area of cylinders.

To determine the amount of material needed to make the aluminum can in figure (a) below, we need to find its total surface area—the sum of the areas of its bases and its lateral surface area.

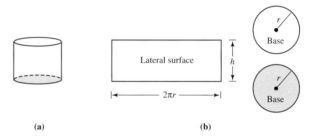

(a)   (b)

To derive a formula for the total surface area of a right circular cylinder, we will take an approach similar to the one we used with prisms. In figure (b), the can is disassembled, and the pieces of aluminum are laid out flat. We observe the following.

- The area of one of the circular bases is πr^2. Therefore, the sum of the areas of two circular bases is $\pi r^2 + \pi r^2 = 2\pi r^2$.

- When the lateral surface of the can is "unrolled," its shape is rectangular. The area of the rectangle is the product of its length $2\pi r$ (the circumference of one circular base) and the height h of the can: $LSA = 2\pi rh$.

These observations suggest the following formula.

Total surface area of a right circular cylinder

The total surface area *TSA* of a right circular cylinder is the sum of the area of its two circular bases and its lateral surface area. If h is the height and r the radius of the base, then the total surface area of a right circular cone is given by

$$TSA = 2\pi r^2 + 2\pi rh$$

EXAMPLE 3   Find the total surface area of the right circular cylinder shown on the right. Give the exact answer and an approximation to the nearest tenth.

16 mm   47 mm

Strategy
We will substitute values for r and h in the formula $TSA = 2\pi r^2 + 2\pi rh$ and evaluate the right side.

Why
The notation *TSA* represents the unknown total surface area of the cylinder.

Self Check 3
Find the total surface area of the right circular cylinder shown below. Give the exact answer and an approximation to the nearest tenth.

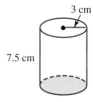

3 cm

7.5 cm

Solution

In this position, one of the cylinder's circular bases is facing the reader. Since its diameter is 16 millimeters, the radius r is 8 mm. If the cylinder were standing vertically, its height h would be 47 mm. To find the total surface area, we proceed as follows.

$$TSA = 2\pi r^2 + 2\pi rh$$ This is the formula for the total surface area of a cylinder.

$$TSA = 2\pi(8)^2 + 2\pi(8)(47)$$ Substitute 8 for r, the radius of the base, and 47 for h, the height of the cylinder.

$$= 2\pi(64) + 2\pi(376)$$

$$= 128\pi + 752\pi$$ Do each multiplication. The area of the bases is 128π mm^2, and the *LSA* is 752π mm^2.

$$= 880\pi$$ Combine like terms.

The total surface area is exactly 880π mm^2. To the nearest tenth, this is 2,764.6 mm^2.

Answer: 63π cm$^2 \approx 197.9$ cm^2

Now Try
Problem 35

3. Find the total surface area of pyramids.

Recall that a *right regular pyramid* is a pyramid whose base is a regular polygon and whose vertex is equidistant from each vertex of the base. Figure (a) below shows a regular pyramid—more specifically, a right regular square pyramid. The four lateral faces of the pyramid are congruent triangles. The height s of each lateral face is called the slant height of the pyramid.

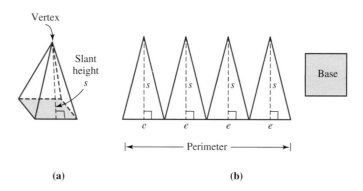

(a)   (b)

To derive a formula for the total surface area of a right regular pyramid, we can "unfold" the figure and lay the pieces out flat, as shown in figure (b). If the length of each base edge is e, then the area of each triangle is one-half of the product of the length of its base and the slant height, or $\frac{1}{2}es$. To find the lateral surface area of the pyramid, we need to find the sum of areas of the four triangular lateral faces.

$$LSA = \frac{1}{2}es + \frac{1}{2}es + \frac{1}{2}es + \frac{1}{2}es$$

$$LSA = \frac{1}{2}s(e + e + e + e)$$ Factor out the common factor of $\frac{1}{2}s$.

The expression $e + e + e + e$ is simply the sum of the lengths of the four edges of the base. We can replace it with the variable p, where p is the perimeter of the base.

$$LSA = \frac{1}{2}sp$$ Because $e + e + e + e = p$.

$$LSA = \frac{1}{2}ps$$ Apply the commutative property of multiplication to write the variable factors in alphabetical order.

These observations suggest the following formula.

Total surface area of a right regular pyramid

> The total surface area *TSA* of a right regular pyramid is the sum of the area of its base and its lateral surface area. If *B* is the area of the base, *p* the perimeter of the base, and *s* the slant height, then the total surface area of a right regular pyramid is given by
>
> $$TSA = B + \frac{1}{2}ps$$

EXAMPLE 4   Find the total surface area of the right regular pyramid shown in figure (a) below, if its base is an equilateral triangle with sides 8 inches long and its slant height is 11 inches. Give the exact answer and an approximation to the nearest tenth.

The base of the pyramid

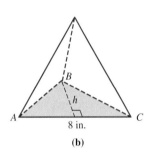

(a)

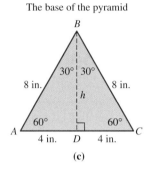

(b)

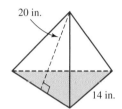

(c)

Self Check 4

Find the total surface area of the right regular pyramid shown below, if its base is an equilateral triangle with sides 14 inches long and its slant height is 20 inches. Give the exact answer and an approximation to the nearest tenth.

Strategy

We will use the properties of a 30°–60°–90° triangle to find the area of the equilateral triangle that is the base of the pyramid.

Why

To find the total surface area of the pyramid using the formula $TSA = B + \frac{1}{2}ps$, we need to know *B*, the area of its base.

Solution

Figures (b) and (c) show the triangular base of the pyramid, which we will call △*ABC*. Because △*ABC* is equilateral, each side is 8 inches long, and each angle has measure 60°. To find its height *h*, we construct altitude $\overline{BD}$ that divides the equilateral triangle into two congruent 30°–60°–90° triangles. Since $\overline{BD}$ is the longer leg and $\overline{DC}$ is the shorter leg of △*DBC*, and since m($\overline{DC}$) = 4 inches, it follows that $h = 4\sqrt{3}$ inches. Knowing *h*, we can now find the area of △*ABC*.

$$\text{Area of } \triangle ABC = \frac{1}{2}bh$$

$$= \frac{1}{2}(8)(4\sqrt{3})$$

$$= 4(4\sqrt{3})$$

$$= 16\sqrt{3}$$

The area of the base of the pyramid is exactly $16\sqrt{3}$ in.².

To find the total surface area of the pyramid, we note that the area of the base is $16\sqrt{3}$ in.², the perimeter of the base is 8 in. + 8 in. + 8 in. = 24 in., and the slant height is 11 in.

$$TSA = B + \frac{1}{2}ps$$ This is the formula for the total surface area of a pyramid.

$$TSA = 16\sqrt{3} + \frac{1}{2}(24)(11)$$ Substitute $16\sqrt{3}$ for *B*, the area of the base, 24 for *p*, the perimeter of the base, and 11 for *s*, the slant height.

$= 16\sqrt{3} + 12(11)$ Multiply $\frac{1}{2}(24)$.

$= 16\sqrt{3} + 132$ Complete the multiplication. The area of the base is $16\sqrt{3}$ in.², and the *LSA* is 132 in.².

Answer: $(49\sqrt{3} + 420)$ in.²
≈ 504.9 in.²

The total surface area of the pyramid is exactly $(16\sqrt{3} + 132)$ in.². To the nearest tenth, this is 159.7 in.².

Now Try
Problem 39 ■

4. Find the total surface area of cones.

Figure (a) below shows a right circular cone. Right circular cones have both a height and a **slant height.** The slant height s is the length of a line segment that joins the vertex of the cone to a point on the edge of its circular base.

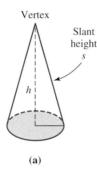
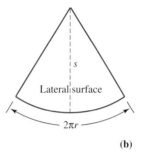

To derive the formula for the total surface area of a right circular cone, we disassemble it and lay the pieces out flat, as shown in figure (b). We observe the following.

* The area of its circular base is πr^2.

* When the lateral surface of the can is "unrolled," a figure called a *sector* results. This sector is almost triangular-shaped. The length of the base of this "triangle" is the circumference of the circular base of the cone. The height of this "triangle" is the slant height s of the cone. These observations suggest that the area of the sector is $\frac{1}{2}bh = \frac{1}{2}(2\pi r)s = \pi rs$.

Total surface area of a right circular cone

> The total surface area *TSA* of a right circular cone is the sum of the area of its base and its lateral surface area. If r is the radius of the base and s the slant height, then the total surface area of a right circular cone is given by
>
> $$TSA = \pi r^2 + \pi rs$$

EXAMPLE 5   Find the total surface area of the right circular cone shown on the right. Give the exact answer and an approximation to the nearest tenth.

Strategy
We will use the height of the cone and its radius to find the slant height.

Why
To find the total surface area of the cone using the formula $TSA = \pi r^2 + \pi rs$, we need to know s, its slant height.

Solution
From the figure, we see that the segments representing the height, a radius, and a slant height form a right triangle. We can use the Pythagorean theorem to find the unknown slant height s of the cone.

$a^2 + b^2 = c^2$ This is the Pythagorean equation.
$12^2 + 9^2 = s^2$ Substitute 12 for a, 9 for b, and s for c.

Self Check 5
Find the total surface area of the right circular cone shown below. Give the exact answer and an approximation to the nearest tenth.

$144 + 81 = s^2$ Evaluate the exponential expressions.

$225 = s^2$ Do the addition.

By the square root property, $s = \sqrt{225} = 15$ or $s = -\sqrt{225} = -15$. Since the slant height must be positive, we have

$s = 15$

The slant height of the cone is 15 cm.

To find the total surface area of the cone, we proceed as follows.

$TSA = \pi r^2 + \pi rs$	This is the formula for the surface area of a right circular cone.
$TSA = \pi(9)^2 + \pi(9)(15)$	Substitute: The radius r of the base is 9 cm, and the slant height s is 15.
$= 81\pi + 135\pi$	The area of the base of the cone is 81π cm^2, and its *LSA* is 135π cm^2.
$= 216\pi$	Combine like terms.

The total surface of the cone is exactly 216π cm^2. To the nearest tenth, this is 678.6 cm^2.

Answer: 96π m$^2 \approx 301.6$ m^2

Now Try
Problem 43

5. Find the surface area of spheres.

We can think of a sphere as a hollow ball. More formally, a sphere is the set of all points that lie a fixed distance r from a point called the *center*. A segment drawn from the center of the sphere to a point on the sphere is called a *radius*. There is a formula to find the surface area of a sphere.

Surface area of a sphere

> The surface area *SA* of a sphere with radius r is given by
>
> $$SA = 4\pi r^2$$

EXAMPLE 6   *Manufacturing beach balls.*   A beach ball is to have a diameter of 16 inches. How many square inches of material will be needed to make the ball? (Disregard any waste.)

Strategy
We will find the radius of the beach ball and substitute that value of r into the formula $SA = 4\pi r^2$.

Why
The amount of material that is needed to make the beach ball is equal to its surface area.

Solution
Since a radius r of the beach ball is one-half the diameter, $r = 8$ inches. To find the surface area of the ball, we proceed as follows.

$SA = 4\pi r^2$	This is the formula for the surface area of a sphere.
$SA = 4\pi(8)^2$	Substitute 8 for r, the radius of the sphere.
$SA = 4\pi(64)$	Evaluate the exponential expression.
$SA = 256\pi$	Multiply: $4 \cdot 64 = 256$.
≈ 804.2477193	Use a calculator.

Exactly 256π in.^{2} of material are needed to make the beach ball. This is a little more than 804 in.^{2}.

Self Check 6
Find the surface area of a beach ball with a diameter twice that of the ball shown in Example 6.

Answer: $1{,}024\pi$ in.$^2 \approx$ 3,217 in.^{2}

Now Try
Problem 47

STUDY SET Section 11

VOCABULARY *Fill in the blanks.*

1. The total _____ of a right prism is the sum of the area of its bases and its lateral faces.

2. Copy the figure below. Label the vertex, the slant height *s*, and the height *h* of the right square pyramid.

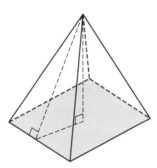

3. The _____ surface area of a pyramid is the sum of the area of its triangular lateral faces.

4. A right _____ pyramid is a pyramid whose base is a regular polygon and whose vertex is equidistant from each vertex of the base.

5. The total surface area of a right circular cylinder is the sum of the area of its two circular _____ and its lateral surface area.

6. Copy the figure below. Label the vertex, the slant height *s*, the height *h* of the right circular cone, and the radius *r* of the base of the cone.

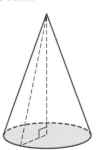

7. If a segment from the vertex of a cone to the center of the base of the cone is perpendicular to the base, the cone is called a _____ circular cone.

8. A _____ is the set of all points that lie a fixed distance *r* from a point called the center.

CONCEPTS

9. Which of the following are acceptable units of measurement for surface area?

ft^2	mi^3	seconds
cubic inches	mm	square yards
gallons	cm^2	meters

10. Suppose the area of the base of a pyramid is 55 ft^2 and its lateral surface area is 144 ft^2. What is the total surface area of the figure?

11. The figure below shows a right rectangular prism that has been "disassembled" and laid out flat. Fill in the blanks.

a. When we find the area of the blue-shaded regions, we are finding the area of the _____.

b. When we find the area of the red-shaded regions, we are finding the _____ surface area.

12. Draw a picture of each figure after it has been "disassembled" and laid out flat. Label the bases and the lateral surface area.

a. cube   **b.** right circular cone

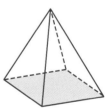

c. right circular cylinder **d.** right square pyramid

13. What is the perimeter *p* of the base of the right regular hexagonal prism shown below?

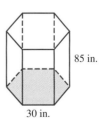

85 in.

30 in.

14. Give the formula that can be used to find the area *B* of the base of each figure.

    a. right square prism   **b.** right triangular pyramid

 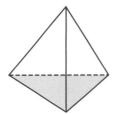

    c. right rectangular prism **d.** right circular cone

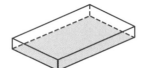 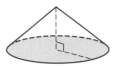

15. Complete the table.

Figure	Total Surface Area Formula
Right prism	
Right circular cylinder	
Right regular pyramid	
Right circular cone	
Sphere	

16. Refer to the figure below. Find *r*.

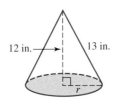

17. Refer to the equilateral triangle shown below.
    a. Find *h*.
    b. Find the area of the triangle. Give the exact answer and an approximation to the nearest hundredth.

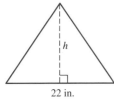

18. Which geometric concept (perimeter, circumference, area, volume, or surface area) should be applied to find each of the following?
    a. The size of a room to be air conditioned
    b. The amount of floor space in a store

    c. The amount of space in a refrigerator freezer
    d. The amount of cardboard in a shoe box
    e. The distance around a checkerboard
    f. The amount of material used to make a basketball

19. a. Simplify: $19\pi + 22\pi$
    b. Multiply: $\frac{1}{2}(11)(8)$

20. Evaluate $4\pi r^2$ for $r = 6$. Give the exact answer and an approximation to the nearest hundredth.

NOTATION

21. In the formula $TSA = 2B + hp$, give what is represented by each of the following.
    a. *TSA*   **b.** *B*
    c. *h*   **d.** *p*

22. In the formula $TSA = B + \frac{1}{2}ps$, give what is represented by each of the following.
    a. *TSA*   **b.** *B*
    c. *p*   **d.** *s*

GUIDED PRACTICE

Find the total surface area of each right prism. The measurements are in inches. **See Example 1.**

23.

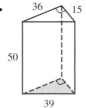

24.

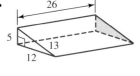

25.

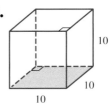

26.

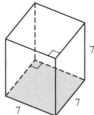

27.

28.

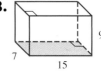

29.

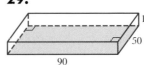

30.

The total surface area of each right regular prism is given. Find the area of one base. **See Example 2.**

31. *TSA* = 1,258 cm^2

32. *TSA* = 1,646 ft^2

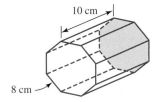

10 cm

8 cm

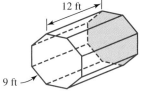

12 ft

9 ft

33. *TSA* = 1,396 in.^{2}

34. *TSA* = 370 yd^2

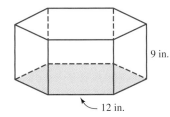

9 in.

12 in.

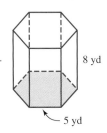

8 yd

5 yd

Find the total surface area of each right circular cylinder. Give the exact answer and an approximation to the nearest tenth. The measurements are in feet. **See Example 3.**

35.

5

8

36.

16

29.5

37.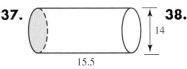

15.5

38.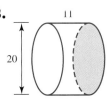

14

11

20

Find the area of the base of the right regular pyramid. Then find its total surface area. Give the exact answer and an approximation to the nearest tenth. The measurements are in meters. **See Example 4.**

39.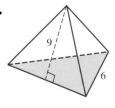

9

6

40.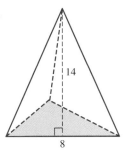

14

8

41.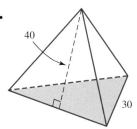

40

30

42.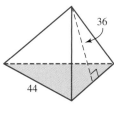

36

44

Find the total surface area of each right circular cone. Give the exact answer and an approximation to the nearest tenth. The measurements are in inches. **See Example 5.**

43.

24

7

44.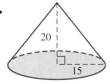

20

15

45.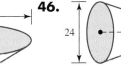

40

21

46.

24

35

Find the surface area of each sphere. Give the exact answer and an approximation to the nearest tenth. The measurements are in inches. **See Example 6.**

47.

3

48.

17

49.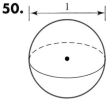

15

50.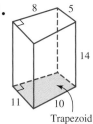

1

TRY IT YOURSELF

Find the total surface area of each right trapezoidal prism. The measurements are in feet.

51.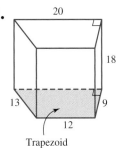

8   5

14

11   10

Trapezoid

52.

20

18

13   9

12

Trapezoid

The area of one base of a right regular hexagonal prism is given. Find its total surface area. Give the exact answer and an approximation to the nearest tenth.

53.

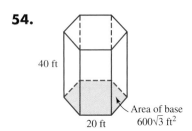

Area of base
$6\sqrt{3}$ ft^2

6 ft

2 ft

54.

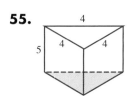

40 ft

20 ft

Area of base
$600\sqrt{3}$ ft^2

The base of each right prism is an equilateral triangle. Find the area of one base of the prism. Then find its total surface area. Give the exact answer and an approximation to the nearest tenth. The measurements are in centimeters.

55.

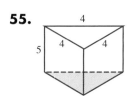

4

5   4   4

56.

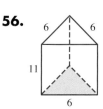

6   6

11

6

Find the surface area of each right square pyramid. The measurements are in yards.

57.

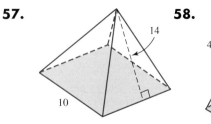

14

10

58.

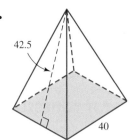

42.5

40

59.

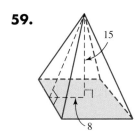

15

8

60.

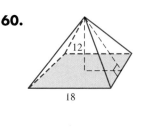

12

18

Find the area of the base of the right regular pyramid. Then find its total surface area. Give the exact answer and an approximation to the nearest tenth. The measurements are in meters.

61.

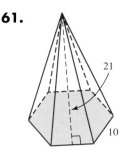

21

10

62.

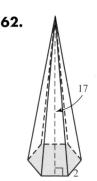

17

2

63. The total surface area of a right square prism is 230 ft^2. Find the height of the prism if the length of a side of a base is 5 ft.

64. The total surface area of a right circular cylinder is 96π in.^{2}. If the radius of its base is 6 in., what is its height?

65. The lateral surface area of a regular square pyramid is 240 m^2. If each side of its base is 10 m long, what is its slant height?

66. The area of the base of a right circular cone is 400π cm^2. If the slant height is 29 cm, find the height of the cone.

67. If the lateral surface area of a right circular cylinder is 42π ft^2 and the total surface area is 60π ft^2, what is the radius of a base?

68. The surface area of a sphere is 196π yd^2. How long is its radius?

69. The height of a cone is 24 inches, and the radius of the base is 7 inches long. Find the lateral surface area of the cone. Give the exact answer and an approximation to the nearest tenth.

70. The total surface area of a cube is 24 cm^2. Find its volume.

71. Find the total surface area of a right prism whose bases are regular hexagons with sides 12 millimeters long and whose height is 10 mm. Give the exact answer and an approximation to the nearest tenth.

72. The base of a right prism is a rhombus with diagonals measuring 10 feet and 24 feet. The height of the prism is 60 feet. Find the total surface area of the figure.

APPLICATIONS

73. PENCILS   Determine the lateral surface area of the pencil shown below.

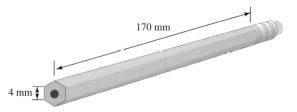

170 mm

4 mm

74. LIGHT   A triangular prism separates white light into a visible spectrum composed of primary colors. In the figure below, the height of the prism is 10 cm, and each edge of its base is 2 cm long. Find the total surface area. Give the exact answer and an approximation to the nearest tenth.

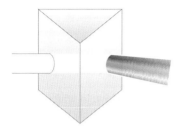

75. WASHINGTON, D.C.   The Washington Monument is a tall shaft of marble blocks that is topped by a regular square pyramid. To find its surface area, we can "disassemble" it as shown below. Use the given information to estimate the lateral surface area of the monument. (In this case, do not include the area of its base.)

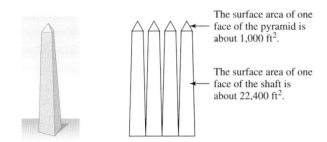

The surface area of one face of the pyramid is about 1,000 ft².

The surface area of one face of the shaft is about 22,400 ft².

76. LINT REMOVER   The illustration below shows a handy gadget; it uses a cylinder of sheets of sticky paper that can be rolled over clothing and furniture to pick up lint and pet hair. After the paper is full, that sheet is peeled away to expose another sheet of sticky paper. Find the area of the first sheet. Give the exact answer and an approximation to the nearest tenth.

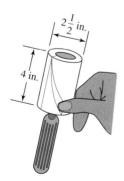

77. ROCKETRY   The model rocket nose cone shown below slips into one end of a cardboard tube that has an outside diameter of 2.4 inches. Find the lateral surface area of the nose cone. Give the exact answer and an approximation to the nearest square inch.

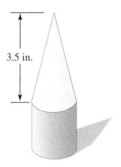

3.5 in.

78. BOWLING   A bowling ball is packaged within a tightly fitting cubical cardboard box, as shown below.
a. Find the amount of cardboard needed to make the box.
b. Find the surface area of the ball. Give the exact answer and an approximation to the nearest square inch.

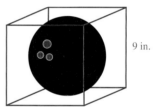

9 in.

WRITING

79. What is meant by *lateral surface area* of a figure? Give an example.

80. Explain how the Pythagorean theorem is used in this section.

81. Explain the difference between the *surface area* and the *volume* of a figure.

82. A right rectangular prism has length 2 ft, width 3 ft, and height 4 ft. Suppose each dimension is then doubled. How much greater will the surface area of the new prism be compared to that of the original prism? Explain how you arrive at your answer.

Sections 8-11 Cumulative Review Problems

Match each concept in Column I with the best response in Column II.

1. Pythagorean theorem

a. $\dfrac{a}{b} = \dfrac{c}{d}$

2. congruent triangles

b. Corresponding sides are proportional.

3. volume

c. $45° - 45° - 90°$ triangle

4. proportion

d.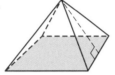

5. right isosceles triangle

e. six pairs of congruent corresponding parts

6. $30°-60°-90°$ triangle

f. lateral surface area

7. similar triangles

g. capacity

8. prism

h. $a^2 + b^2 = c^2$

9. LSA

i. shorter leg half as long as hypotenuse

10. slant height

j.

11. See the figure below, where $\triangle ABC \cong \triangle DEF$. Name the six corresponding parts of the congruent triangles.

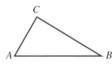 

$\angle A \cong$ _____

$\angle B \cong$ _____

$\angle C \cong$ _____

$\overline{AC} \cong$ _____

$\overline{AB} \cong$ _____

$\overline{BC} \cong$ _____

12. Tell whether each pair of triangles are congruent. If they are, tell why.

a.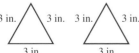

b.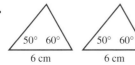

c.

d.

13. Refer to the figure below, in which $\triangle ABC \cong \triangle DEF$.

  a. Find $m(\overline{DE})$.

  b. Find $m(\angle E)$.

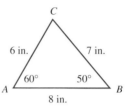 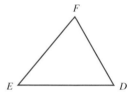

14. Tell whether the triangles in each pair are similar.

a.

b.

15. Refer to the triangles below. Find x and y. The units are meters.

  a. Find x.

  b. Find y.

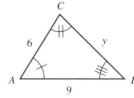 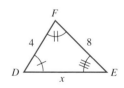

16. SHADOWS If a tree casts a 7-foot shadow at the same time as a man 6 feet tall casts a 2-foot shadow, how tall is the tree?

17. Refer to the right triangle below. Find the missing side length. Approximate any exact answers that contain a radical to one decimal place.

  a. Find c if $a = 10$ cm and $b = 24$ cm.

  b. Find b if $a = 6$ in. and $c = 8$ in.

18. Find the missing lengths of each triangle. Give the exact answer and an approximation to the nearest hundredth if an answer contains a radical.

a.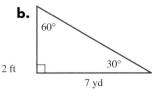

b.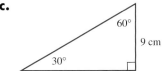

c.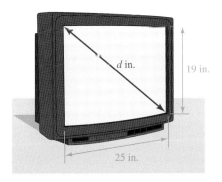

19. TELEVISION   To the nearest tenth of an inch, what is the diagonal measurement of the television screen shown below?

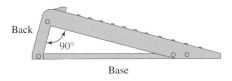

20. CAR REPAIR   To create some space to work under the front end of a car, a mechanic drives it up steel ramps. See the figure below. The ramp is 1 foot longer than the back, and the base is 2 feet longer than the back of the ramp. Find the length of each side of the ramp.

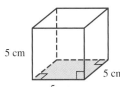

21. How many cubic inches are there in 1 cubic foot?

22. Find the volume of each figure. Give the exact answer and an approximation to the nearest hundredth if an answer contains π.

a.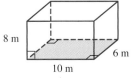

b.

c.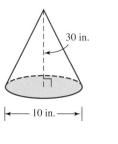

d.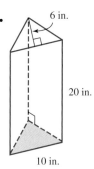

e.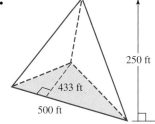

23. FARMING   A silo is used to store wheat and corn. Find the volume of the silo shown below. Give the exact answer and an approximation to the nearest cubic foot.

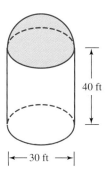

24. A square pyramid has a volume of 48 in.^{3}. Find its height if the length of a side of its base is 6 inches.

25. Find the total surface area of each figure. Give the exact answer and an approximation to the nearest hundredth if an answer contains π.

a.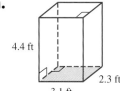

b.

c.

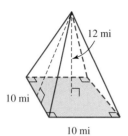

12 mi

10 mi

10 mi

d.

5 in.

26. The surface area of a right circular cone is 880π ft^2. The radius of its circular base is 20 feet. What is its slant height?

27. Give a real-life example in which the concept of perimeter is used. Do the same for area and for volume. Be sure to discuss the type of units used in each case.

28. What is the difference between lateral surface area and total surface area of a pyramid? Make a drawing to help explain your answer.

Objectives

1. Use inductive reasoning to solve problems.
2. Use deductive reasoning to solve problems.

To reason means to think logically. The objective of this appendix is to develop your problem-solving ability by improving your reasoning skills. We will introduce two fundamental types of reasoning that can be applied in a wide variety of settings. They are known as *inductive reasoning* and *deductive reasoning.*

© tompiodesign.com / Alamy

1. Use inductive reasoning to solve problems.

In a laboratory, scientists conduct experiments and observe outcomes. After several repetitions with similar outcomes, the scientist will generalize the results into a statement that appears to be true:

- If I heat water to 212°F, it will boil.
- If I drop a weight, it will fall.
- If I combine an acid with a base, a chemical reaction occurs.

When we draw general conclusions from specific observations, we are using **inductive reasoning.** The next examples show how inductive reasoning can be used in mathematical thinking. Given a list of numbers or symbols, called a *sequence,* we can often find a missing term of the sequence by looking for patterns and applying inductive reasoning.

EXAMPLE 1   Find the next number in the sequence 5, 8, 11, 14, ...

Strategy
We will find the *difference* between pairs of numbers in the sequence.

Why
This process will help us discover a pattern that we can use to find the next number in the sequence.

Solution
The numbers in the sequence 5, 8, 11, 14, . . . are increasing. We can find the difference between each pair of successive numbers as follows:

Self Check 1
Find the next number in the sequence −3, −1, 1, 3, ...

$8 - 5 = 3$ Subtract the first number, 5, from the second number, 8.

$11 - 8 = 3$ Subtract the second number, 8, from the third number, 11.

$14 - 11 = 3$ Subtract the third number, 11 from the fourth number, 14.

The difference between each pair of numbers is 3. This means that each number in the sequence is 3 greater than the previous one. Thus, the next number in the sequence is $14 + 3$, or 17.

Answer: 5

Now Try
Problem 11

EXAMPLE 2   Find the next number in the sequence $-2, -4, -6, -8, \ldots$.

Strategy
The terms of the sequence are decreasing. We will determine how each number differs from the previous number.

Why
This type of examination helps us discover a pattern that we can use to find the next number in the sequence.

Solution

This number is 2 less than the previous number.	This number is 2 less than the previous number.	This number is 2 less than the previous number.

-2 , -4 , -6 , -8 , $\ldots$

Since each successive number is 2 less than the previous one, the next number in the sequence is $-8 - 2$, or -10.

Self Check 2
Find the next number in the sequence $-0.1, -0.3, -0.5, -0.7, \ldots$.

Answer: -0.9

Now Try
Problem 15

EXAMPLE 3   Find the next letter in the sequence A, D, B, E, C, F, D, ...

Strategy
We will create a letter–number correspondence and rewrite the sequence in an equivalent numerical form.

Why
Many times, it is easier to determine the pattern if we examine a sequence of numbers instead of letters.

Solution
The letter A is the 1st letter of the alphabet, D is the 4th letter, B is the 2nd letter, and so on. We can create the following letter–number correspondence:

Letter   Number

A → 1   ⎱
D → 4   ⎰ Add 3.
B → 2   ⎰ Subtract 2.
E → 5   ⎱ Add 3.
C → 3   ⎰ Subtract 2.
F → 6   ⎱ Add 3.
D → 4   ⎰ Subtract 2.

The numbers in the sequence 1, 4, 2, 5, 3, 6, 4, . . . alternate in size. They change from smaller to larger, to smaller, to larger, and so on.

We see that 3 is added to the first number to get the second number. Then 2 is subtracted from the second number to get the third number. To get successive numbers in the sequence, we alternately add 3 to one number and then subtract 2 from that result to get the next number.

Applying this pattern, the next number in the given numerical sequence would be $4 + 3$, or 7. The next letter in the original sequence would be G, because it is the 7th letter of the alphabet.

Self Check 3
Find the next letter in the sequence B, G, D, I, F, K, H, ...

Answer: M

Now Try
Problem 19

EXAMPLE 4   Find the next shape in the sequence below.

Strategy
To find the next shape in the sequence, we will focus on the changing positions of the dots.

Why
The star does not change in any way from term to term.

Solution
We see that each of the three dots moves from one point of the star to the next, in a counterclockwise direction. This is a circular pattern. The next shape in the sequence will be the one shown below.

EXAMPLE 5   Find the next shape in the sequence below.

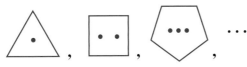

Strategy
To find the next shape in the sequence, we must consider two changing patterns at the same time.

Why
The shapes are changing and the number of dots within them are changing.

Solution
The first figure has three sides and one dot, the second figure has four sides and two dots, and the third figure has five sides and three dots. Thus, we would expect the next figure to have six sides and four dots, as shown below.

Self Check 4
Find the next shape in the sequence below.

Answer:

Now Try
Problem 23   ■

Self Check 5
Find the next shape in the sequence below.

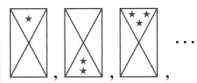

Answer:

Now Try
Problem 27   ■

2. Use deductive reasoning to solve problems.

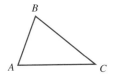

As opposed to inductive reasoning, deductive reasoning moves from the general case to the specific. For example, if we know that the sum of the angles in any triangle is 180°, we know that the sum of the angles of △*ABC* shown in the left margin is 180°. Whenever we apply a general principle to a particular instance, we are using deductive reasoning.

A deductive reasoning system is built on four elements:

1. **Undefined terms:** terms that we accept without giving them formal meaning

2. **Defined terms:** terms that we define in a formal way

3. **Axioms** or **postulates:** statements that we accept without proof

4. **Theorems:** statements that we can prove with formal reasoning

Many problems can be solved by deductive reasoning. For example, suppose a student knows that his college offers algebra classes in the morning, afternoon, and evening and that Professors Anderson, Medrano, and Ling are the only algebra instructors at the school. Furthermore, suppose that the student plans to enroll in a morning algebra class. After some investigating, he finds out that Professor Anderson teaches only in the afternoon and Professor Ling teaches only in the evening. Without knowing anything about Professor Medrano, he can conclude that she will be his algebra teacher, since she is the only remaining possibility.

The following examples show how to use deductive reasoning to solve problems.

EXAMPLE 6 *Scheduling classes.* An online college offers only one calculus course, one algebra course, one statistics course, and one trigonometry course. Each course is to be taught by a different professor. The four professors who will teach these courses have the following course preferences:

1. Professors A and B don't want to teach calculus.

2. Professor C wants to teach statistics.

3. Professor B wants to teach algebra.

Who will teach trigonometry?

Strategy
We will construct a table showing all the possible teaching assignments. Then we will cross off those classes that the professors do not want to teach.

Why
The best way to examine this much information is to describe the situation using a table.

Solution
The following table shows each course, with each possible instructor.

Calculus	Algebra	Statistics	Trigonometry
A	A	A	A
B	B	B	B
C	C	C	C
D	D	D	D

Since Professors A and B don't want to teach calculus, we can cross them off the calculus list. Since Professor C wants to teach statistics, we can cross her off every other list. This leaves Professor D as the only person to teach calculus, so we can cross her off every other list. Since Professor B wants to teach algebra, we can cross him off every other list. Thus, the only remaining person left to teach trigonometry is Professor A.

Calculus	Algebra	Statistics	Trigonometry
A̶	A	A	A
B̶	B	B̶	B̶
C̶	C̶	C	C̶
D	D̶	D̶	D̶

Now Try
Problem 31

EXAMPLE 7 *State flags.* The graph below gives the number of state flags that feature an eagle, a star, or both. How many state flags have neither an eagle nor a star?

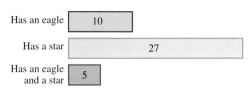

FLAG OF NORTH DAKOTA

Strategy
We will use two intersecting circles to model this situation.

Why
The intersection is a way to represent the number of state flags that have both an eagle and a star.

Solution
In figure (a) below, the intersection (overlap) of the circles shows that there are 5 state flags that have both an eagle and a star. If an eagle appears on a total of 10 flags, then the red circle must contain 5 more flags outside of the intersection, as shown in figure (b). If a total of 27 flags have a star, the blue circle must contain 22 more flags outside the intersection, as shown.

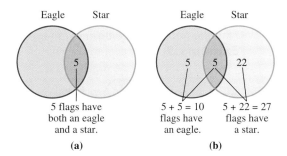

From figure (a), we see that $5 + 5 + 22$, or 32 flags have an eagle, a star, or both. To find how many flags have neither an eagle nor a star, we subtract this total from the number of state flags, which is 50.

$$50 - 32 = 18$$

There are 18 state flags that have neither an eagle nor a star.

Self Check 7
Of the 50 cars on a used-car lot, 9 are red, 31 are foreign models, and 6 are red, foreign models. If a customer wants to buy an American model that is not red, how many cars does she have to choose from?

Answer: 16

Now Try
Problem 35

STUDY SET   Appendix I

VOCABULARY   *Fill in the blanks.*

1. _____ reasoning draws general conclusions from specific observations.

2. _____ reasoning moves from the general case to the specific.

CONCEPTS   *Tell whether the pattern shown is increasing, decreasing, alternating, or circular.*

3. 2, 3, 4, 2, 3, 4, 2, 3, 4, . . .

4. 8, 5, 2, −1, . . .

5. −2, −4, 2, 0, 6, . . .

6. 0.1, 0.5, 0.9, 1.3, . . .

7. a, c, b, d, c, e, . . .

8. 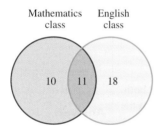 . . .

9. ROOM SCHEDULING From the chart, determine what time(s) on a Wednesday morning a practice room in a music building is available. The symbol X indicates that the room has already been reserved.

	M	T	W	Th	F
9 A.M.	X		X		X
10 A.M.	X	X			X
11 A.M.		X	X	X	

10. COUNSELING QUESTIONNAIRE   A group of college students were asked whether they were taking a mathematics course and whether they were taking an English course. The results are displayed below.

a. How many students were taking a mathematics course and an English course?

b. How many students were taking an English course but not a mathematics course?

c. How many students were taking a mathematics course?

Mathematics class — English class

10   11   18

GUIDED PRACTICE   *Find the number that comes next in each sequence.* **See Example 1.**

11. 1, 5, 9, 13, . . .

12. 11, 20, 29, 38, . . .

13. 5, 9, 14, 20, . . .

14. 6, 8, 12, 18, . . .

Find the number that comes next in each sequence. **See Example 2.**

15. 15, 12, 9, 6, . . .

16. 81, 77, 73, 69, . . .

17. −3, −5, −8, −12, . . .

18. 1, −8, −16, −25, −33, . . .

Find the letter that comes next in each sequence. **See Example 3.**

19. E, I, H, L, K, O, N, . . .

20. C, H, D, I, E, J, F, . . .

21. c, b, d, c, e, d, f, . . .

22. z, w, y, v, x, u, w, . . .

Find the figure that comes next in each sequence. **See Example 4.**

23.  . . .

24.  . . .

25. . . .

26. . . .

Find the figure that comes next in each sequence. **See Example 5.**

27.  . . .

28.  . . .

29. 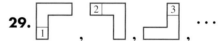 . . .

30.  . . .

What conclusion can be drawn from each set of information? **See Example 6.**

31. TEACHING SCHEDULES   A small college offers only one biology course, one physics course, one chemistry course, and one zoology course. Each course is to be taught by a different adjunct professor. The four professors who will teach these courses have the following course preferences:

1. Professors B and D don't want to teach zoology.

2. Professor A wants to teach biology.

3. Professor B wants to teach physics.

Who will teach chemistry?

32. DISPLAYS   Four companies will be displaying their products on tables at a convention. Each company will be assigned one of the displays shown below. The companies have expressed the following preferences:

1. Companies A and C don't want display 2.

2. Company A wants display 3.

3. Company D wants display 1.

Which company will get display 4?

Display 1   Display 2   Display 3   Display 4

33. OCUPATIONS   Four people named John, Luis, Maria, and Paula have occupations as teacher, butcher, baker, and candlestick maker.

1. John and Paula are married.

2. The teacher plans to marry the baker in December.

3. Luis is the baker.

Who is the teacher?

34. PARKING   A Ford, a Buick, a Dodge, and a Mercedes are parked side by side.

1. The Ford is between the Mercedes and the Dodge.

2. The Mercedes is not next to the Buick.

3. The Buick is parked on the left end.

Which car is parked on the right end?

Use a circle diagram to solve each problem. **See Example 7.**

35. EMPLOYMENT HISTORY   One hundred office managers were surveyed to determine their employment backgrounds. The survey results are shown below. How many office managers had neither sales nor manufacturing experience?

36. PURCHASING TEXTBOOKS   Sixty college sophomores were surveyed to determine where they purchased their textbooks during their freshman year. The survey results are shown below. How many students did not purchase a book at a bookstore or online?

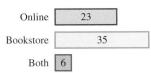

37. SIBLINGS   When 27 children in a first-grade class were asked, "How many of you have a brother?," 11 raised their hands. When asked, "How many have a sister?," 15 raised their hands. Eight children raised their hands both times. How many children didn't raise their hands either time?

38. PETS   When asked about their pets, a group of 35 sixth-graders responded as follows:

- 21 said they had at least one dog.
- 11 said they had at least one cat.
- 5 said they had at least one dog and at least one cat.

How many of the students do not have a dog or a cat?

TRY IT YOURSELF   *Find the next letter or letters in the sequence.*

39. A, c, E, g, . . .   **40.** R, SS, TTT, . . .

41. Z, A, Y, B, X, C, . . .   **42.** B, N, C, N, D, . . .

Find the missing figure in each sequence.

43.

44.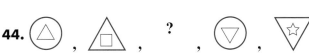

Find the next letter in the sequence.

45. C, B, F, E, I, H, L, . . .

46. d, h, g, k, j, n, . . .

Find the next number in the sequence.

47. −7, 9, −6, 8, −5, 7, −4, . . .

48. 2, 5, 3, 6, 4, 7, 5, . . .

49. 9, 5, 7, 3, 5, 1, . . .

50. 1.3, 1.6, 1.4, 1.7, 1.5, 1.8, . . .

51. −2, −3, −5, −6, −8, −9, . . .

52. 8, 5, 1, −4 , −10 , −17, . . .

53. 6, 8, 9, 7, 9, 10, 8, 10, 11, . . .

54. 10, 8, 7, 11, 9, 8, 12, 10, 9, . . .

55. ZOOS   In a zoo, a zebra, a tiger, a lion, and a monkey are to be placed in four cages numbered from 1 to 4, from left to right. The following decisions have been made:

1. The lion and the tiger should not be side by side.

2. The monkey should be in one of the end cages.

3. The tiger is to be in cage 4.

In which cage is the zebra?

56. FARM ANIMALS   Four animals—a cow, a horse, a pig, and a sheep—are kept in a barn, each in a separate stall.

   1. The cow is in the first stall.

   2. Neither the pig nor the sheep can be next to the cow.

   3. The pig is between the horse and the sheep.

   What animal is in the last stall?

57. OLYMPIC DIVING   Four divers at the Olympics finished first, second, third, and fourth.

   1. Diver B beat diver D.

   2. Diver A placed between divers D and C.

   3. Diver D beat diver C.

   In which order did they finish?

58. FLAGS   A green, a blue, a red, and a yellow flag are hanging on a flagpole.

   1. The only flag between the green and yellow flags is blue.

   2. The red flag is next to the yellow flag.

   3. The green flag is above the red flag.

   What is the order of the flags from top to bottom?

APPLICATIONS

59. JURY DUTY   The results of a jury service questionnaire are shown below. Determine how many of the 20,000 respondents have served on neither a criminal court nor a civil court jury.

Jury Service Questionnaire

997	Served on a criminal court jury
103	Served on a civil court jury
35	Served on both

60. ELECTRONIC POLL   For the Internet poll shown below, the first choice was clicked on 124 times, the second choice was clicked on 27 times, and both the first and second choices were clicked on 19 times. How many times was the third choice, "Neither," clicked on?

Internet Poll	You may vote for more than one.
What would you do if gasoline reached $5.50 a gallon?	⊙ Cut down on driving ⊙ Buy a more fuel-efficient car ⊙ Neither
	Number of people voting | 178

61. THE SOLAR SYSTEM   The graph below shows some important characteristics of the nine planets in our solar system. How many planets are neither rocky nor have moons?

Rocky planets	4
Planets with moons	7
Rocky planets with moons	2

62. WORKING TWO JOBS   Andres, Barry, and Carl each have a completely different pair of jobs from the following list: jeweler, musician, painter, chauffeur, barber, and gardener. Use the facts below to find the two occupations of each man.

   1. The painter bought a ring from the jeweler.

   2. The chauffeur offended the musician by laughing at his mustache.

   3. The chauffeur dated the painter's sister.

   4. Both the musician and the gardener used to go hunting with Andres.

   5. Carl beat both Barry and the painter at monopoly.

   6. Barry owes the gardener $100.

WRITING

63. Describe deductive reasoning and inductive reasoning.

64. Describe a real-life situation in which you might use deductive reasoning.

65. Describe a real-life situation in which you might use inductive reasoning.

66. Write a problem in such a way that the diagram below can be used to solve it.

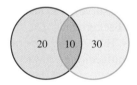

Study Set Section 1 (page 5)

1. point, line, plane **3.** midpoint **5.** angle **7.** protractor
9. right **11.** 180° **13. a.** one **b.** line **15. a.** $\overrightarrow{SR}, \overrightarrow{ST}$
b. S **c.** $\angle RST, \angle TSR, \angle S, \angle 1$
17. a.   **b.**

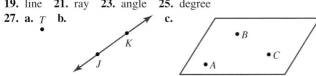

c.   **d.**

19. line **21.** ray **23.** angle **25.** degree
27. a. T **b.**   **c.**

29. 2 **31.** 3 **33.** 1 **35.** 6 **37.** B **39.** D **41.** right
43. acute **45.** straight **47.** obtuse **49.** 50° **51.** 25°
53. 75° **55.** 130° **57.** 180° **59.** 90° **61.** true
63. false. A line does not have an endpoint. **65.** true
67. true **69.** acute **71.** obtuse **73.** right
75. straight **77.** 40° **79.** 135° **81.** 10° **83.** 175°
85.

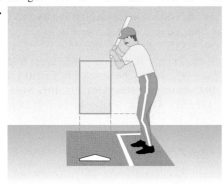

87. a. about 80° **b.** about 30° **c.** about 65°

Study Set Section 2 (page 15)

1. Adjacent **3.** congruent **5.** 90°
7. a.   **b.**

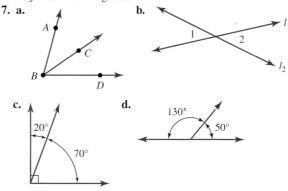

c.   **d.**

9. a. 33° **b.** $2n$ **11.** false **13.** false **15.** yes
17. yes **19.** no **21.** true **23.** true **25.** true **27.** true
29. angle **31.** variable **33.** 10° **35.** 27.5° **37.** 70°
39. 65° **41.** 30°, 60°, 120° **43.** 25°, 115°, 65°
45. 53° **47.** 35° **49.** 60° **51.** 75° **53.** 95° **55.** 41°
57. 50° **59.** 230° **61.** 100° **63.** 40° **65.** 66° **67.** 128°
69. 141° **71.** 1° **73.** 70° **75.** 170° **77.** 65°, 115° **79.** 30°

Study Set Section 3 (page 25)

1. coplanar, noncoplanar **3.** Perpendicular **5.** alternate, corresponding
7. a.   **b.**

9. a.   **b.**

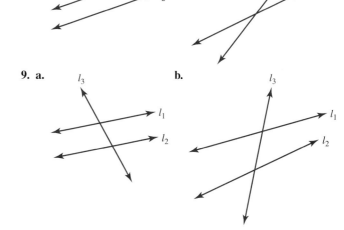

11. corresponding **13.** interior **15.** They are perpendicular.
17. right **19.** perpendicular **21. a.** ∠1 and ∠5, ∠4 and ∠8,
∠2 and ∠6, ∠3 and ∠7 **b.** ∠3, ∠4, ∠5, and ∠6 **c.** ∠3 and ∠5,
∠4 and ∠6 **23.** m(∠1) = 130°, m(∠2) = 50°, m(∠3) = 50°,
m(∠5) = 130°, m(∠6) = 50°, m(∠7) = 50°, m(∠8) = 130°
25. ∠1 ≅ ∠X, ∠2 ≅ ∠N **27.** 12°; 40°, 40° **29.** 10°; 50°, 130°
31. Since a pair of alternate interior angles are congruent,
the lines are parallel. **33. a.** 50°, 135°, 45°, 85° **b.** 180°
c. 180° **35.** Vertical angles: ∠1 ≅ ∠2; alternate interior angles
∠B ≅ ∠D, ∠E ≅ ∠A **37.** 40°, 40°, 140° **39.** 12°, 70°, 70°
41. The plummet string should hang perpendicular to the top of
the stones. **43.** There are perpendicular lines in the center section
of the logo. **45.** The strips of wallpaper should be hung on the
wall parallel to each other and they should be perpendicular to
the floor. **47.** 75°, 105°, 75°

Study Set Section 4 (page 36)

1. polygon **3.** vertex **5.** equilateral, isosceles, scalene
7. hypotenuse, legs **9.** addition
11. a. **b.** **c.**
d. **e.** **f.**

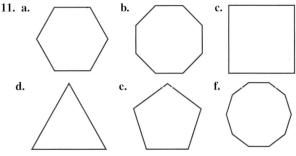

13. a. H, I, J **b.** 3; $\overline{HI}, \overline{IJ}, \overline{JH}$ **c.** △HIJ, △IJH, △JHI
15. a. **b.** **c.**

17. a. 90° **b.** right **c.** $\overline{AB}, \overline{BC}$ **d.** $\overline{AC}$ **e.** $\overline{AC}$ **f.** $\overline{AC}$
19. a. isosceles **b.** converse **21. a.** $\overline{EF} \cong \overline{GF}$
b. isosceles **23.** 180° **25.** triangle **27.** segment
29. $\overline{AB} \cong \overline{CB}$ **31. a.** 4, quadrilateral, 4 **b.** 6, hexagon, 6
33. a. 7, heptagon, 7 **b.** 9, nonagon, 9 **35. a.** scalene
b. isosceles **37. a.** equilateral **b.** scalene **39.** yes
41. no **43.** 55° **45.** 45° **47.** 50°; 50°, 60°, 70°
49. 20°; 20°, 80°, 80° **51.** m(∠A) = 54°, m(∠B) = 42°,
m(∠C) = 84° **53.** m(∠A) = 15°, m(∠B) = 105°, m(∠C) = 60°
55. 68° **57.** 9° **59.** 39° **61.** 44.75° **63.** 28°
65. 73° **67.** 90° **69.** 45° **71.** 90.7° **73.** 61.5° **75.** 12°
77. 52.5° **79.** 39°, 39°, 102° or 70.5°, 70.5°, 39° **81.** 70°, 70°,
40° **83.** 40°, 80°, 60° **85.** m∠A = 100°, m∠B = 45°,
m∠C = 35° **87.** 73° **89.** 75° **91. a.** octagon **b.** triangle
c. pentagon **93.** As the jack is raised, the two sides of the jack
remain the same length. **95.** equilateral triangle

Study Set Section 5 (page 51)

1. quadrilateral **3.** rectangle **5.** rhombus **7.** trapezoid, bases,
isosceles **9. a.** four; A, B, C, D **b.** four; $\overline{AB}, \overline{BC}, \overline{CD}, \overline{DA}$
c. 2; $\overline{AC}, \overline{BD}$ **d.** yes, no, no, yes **11. a.** $\overline{VU}$ **b.** ‖
13. a. right **b.** parallel **c.** length **d.** length **e.** midpoint
15. rectangle **17. a.** no **b.** yes **c.** no **d.** yes **e.** no
f. yes **19. a.** isosceles trapezoid **b.** ∠J, ∠M **c.** ∠K, ∠L
d. m(∠M), m(∠L), m($\overline{ML}$) **21. a.** 7 **b.** 5 **c.** 2
d. $S = (n - 2)180°$ **23.** The four sides of the quadrilateral are the

same length. **25. a.** the sum of the measures of the angles of a
polygon; the number of sides of the polygon **b.** the angle measure of a
regular polygon; the number of sides of the polygon **c.** the measure
of an exterior angle of a regular polygon; the measure of an interior
angle of a regular polygon **d.** the measure of an exterior angle of a
regular polygon; the number of sides of the polygon **27. a.** square,
rhombus, rectangle **b.** trapezoid **c.** rhombus **d.** rectangle
29. a. 90° **b.** 9 **c.** 18 **d.** 18 **31. a.** 42° **b.** 95°
33. a. 9 **b.** 70° **c.** 110° **d.** 110° **35.** 2,160° **37.** 3,240°
39. 1,080° **41.** 1,800° **43.** 5 **45.** 7 **47.** 13 **49.** 14
51. 135° **53.** 108° **55.** 156° **57.** 172.8° **59.** 6 **61.** 20
63. 12 **65.** 50 **67.** 45° **69.** 72° **71.** 24° **73.** 7.2°
75. 20° **77.** 9° **79.** 40° **81.** 60° **83. a.** 30° **b.** 30°
c. 60° **d.** 8 cm **e.** 4 cm **85.** 40°; m(∠A) = 90°,
m(∠B) = 150°, m(∠C) = 40°, m(∠D) = 80° **87. a.** 80 **b.** $4\frac{1}{2}$°
89. 36° **91. a.** trapezoid **b.** square **c.** rectangle **d.** trapezoid **e.** parallelogram **93.** She should measure the lengths of
the diagonals. If they are not the same length, she should adjust the
frame so that they are. **95.** $\frac{3}{4}$ in.; 65°; 115°

Study Set Section 6 (page 67)

1. perimeter **3.** area **5.** area **7.** 128 ft² **9. a.** $P = 4s$
b. $P = 2l + 2w$
11. a. **b.**
c. **d.**

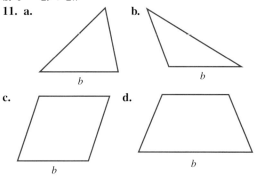

13. a rectangle and a triangle **15. a.** square inch **b.** m²
17. 32 in. **19.** 23 mi **21.** 25.2 m **23.** 31.8 m **25.** 10 ft,
10 ft, 15 ft **27.** lower base: 15 m; upper base: 10 m;
each leg: 5 m **29.** 16 cm² **31.** 6.25 m² **33.** 144
35. 1,000,000 **37.** 27,878,400 **39.** 1,000,000 **41.** 135 ft²
43. 11,160 ft² **45.** 25 in.² **47.** 27 cm² **49.** 7.5 in.²
51. 10.5 mi² **53.** 40 ft² **55.** 91 cm² **57.** 4 m **59.** 12 cm
61. 36 m **63.** 11 mi **65.** 102 in.² **67.** 360 ft² **69.** 75 m²
71. 75 yd² **73.** $1,200 **75.** $4,875 **77.** length 15 in. and width
5 in.; length 16 in. and width 4 in. (Answers may vary.) **79.** sides
of length 5 m **81.** base 5 yd and height 3 yd (Answers may
vary.) **83.** length 5 ft and width 4 ft; length 20 ft and width 3 ft
(Answers may vary.) **85.** 60 cm² **87. a.** $(4x + 4)$ ft
b. $(4x + 6)$ ft **c.** $(3b + 8)$ ft **d.** $15x$ ft **89.** 36 m
91. $28\frac{1}{3}$ ft **93. a.** 9 cm, 21 cm **b.** 189 cm² **95.** upper base:
3 in.; lower base: 13 in. **97.** 36 m **99.** $x = 3.7$ ft; $y = 10.1$ ft;
50.8 ft **101.** 81 **103.** linoleum **105.** $361.20 **107.** $192
109. 111,825 mi² **111.** 51 **113. a.** 3 mi² **b.** 11 + 3; 14 mi²
c. 8.5 mi² **d.** 16; 4 mi²; 16 + 22 = 38; 9.5 mi²; 6.75 mi²

Study Set Section 7 (page 83)

1. radius **3.** diameter **5.** circumference **7.** twice **9.** minor,
major **11.** arc **13.** $\overline{OA}, \overline{OC}$, and $\overline{OB}$ **15.** $\overline{DA}, \overline{DC}$, and $\overline{AC}$
17. $\overparen{ABC}$ and $\overparen{ADC}$ **19.** ∠COB **21. a.** Multiply the radius
by 2. **b.** Divide the diameter by 2. **23.** $C = \pi D$ or $C = 2\pi r$

25. π **27.** square 6   **29.** 12 in.   **31.** $\overparen{YWZ}$
33. 318°   **35.** $\angle YXW$ **37. a.**

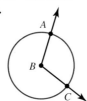

b. $L = \frac{q}{360°} \cdot 2\pi r$ **c.** $2\pi r$ **d.** $\frac{q}{360°}$ **39. a.** 8π **b.** 14π
c. $\frac{25\pi}{3}$ **41.** arc AB **43. a.** multiplication: $2\pi r = 2 \cdot \pi \cdot r$
b. raising to a power and multiplication: $\pi r^2 = \pi \cdot r^2$ **45.** 8π ft
≈ 25.1 ft   **47.** 12π m ≈ 37.7 m   **49.** $\frac{12}{\pi}$ in. ≈ 3.82 in.   **51.** $\frac{9}{\pi}$ cm
≈ 2.86 cm   **53.** 50.85 cm   **55.** 31.42 m   **57.** 9π in.$^2 \approx 28.3$ in.^{2}
59. 81π in.$^2 \approx 254.5$ in.^{2} **61.** 18.9 ft   **63.** 112.3 in.
65. 128.5 cm^2 **67.** 57.1 cm^2 **69.** $\frac{5\pi}{6}$ in. ≈ 2.6 in.   **71.** $\frac{33\pi}{2}$ m $\approx$
51.8 m   **73.** $\frac{25\pi}{3}$ ft$^2 \approx 26.2$ ft^2 **75.** 375π cm$^2 \approx 1{,}178.1$ cm^2
77. a. 1.2 m   **b.** 2.4 m   **c.** 7.4 m   **79.** 27.4 in.^{2} **81.** 66.7 in.^{2}
83. $\frac{3\pi}{2}$ mi ≈ 4.7 mi   **85.** 50π yd ≈ 157.08 yd   **87.** 7 ft
89. 6π in. ≈ 18.8 in.   **91.** $\frac{225\pi}{4}$ in.$^2 \approx 176.7$ in.^{2} **93.** 120π in. $\approx$
377.0 in.   **95.** $\frac{113}{\pi}$ m ≈ 35.97 m   **97.** 20.25π mm$^2 \approx$
63.6 mm^2 **99.** 8 in.   **101.** 5 cm   **103.** 225π ft$^2 \approx 706.86$ ft^2
105. $\frac{5}{2}$ yd $= 2.5$ yd   **107. a.** 1 in.   **b.** 2 in.   **c.** 2π in. ≈ 6.28 in.
d. π in.$^2 \approx 3.14$ in.^{2} **109.** π mi$^2 \approx 3.14$ mi^2
111. $\frac{102.6}{\pi}$ ft ≈ 32.66 ft   **113.** 13 times   **115.** 1.59 ft
117. 4π ft$^2 \approx 12.57$ ft^2; 0.25π ft$^2 \approx 0.79$ ft^2; 6.25%
119. a. $\frac{16\pi}{9}$ in. ≈ 5.59 in.   **b.** $\frac{17\pi}{18}$ in. ≈ 2.97 in.

Cumulative Review Problems   Sections 1–7   (page 90)

1. point, line, plane
2. a.

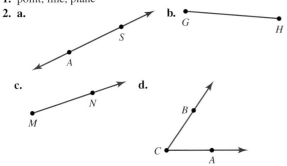

3. no   **4.** $\angle XYZ$, $\angle ZYX$, $\angle Y$, $\angle 1$ **5. a.** 130°, obtuse   **b.** 90°,
right   **c.** 45°, acute   **d.** 180°, straight   **6. a.** measure
b. length   **c.** line   **d.** complementary   **7.** D **8. a.** false
b. true   **c.** true   **d.** true   **e.** false   **9.** 20°; 60°, 60°   **10.** 140°
11. a. transversal   **b.** $\angle 6$ **c.** $\angle 7$ **12.** m($\angle 1$) = 155°,
m($\angle 3$) = 155°, m($\angle 4$) = 25°, m($\angle 5$) = 25°, m($\angle 6$) = 155°,
m($\angle 7$) = 25°, m($\angle 8$) = 155°   **13.** 50°; 110°, 70°   **14.** Yes.
Since a pair of alternate interior angles are congruent, the lines are
parallel.   **15. a.** 8, octagon, 8   **b.** 5, pentagon, 5   **c.** 6,
hexagon, 6   **d.** 4, quadrilateral, 4   **16. a.** isosceles   **b.** scalene
c. equilateral   **d.** isosceles   **17.** 70°   **18.** 70°, 70°, 40°
19. rectangle, square, parallelogram
20. a.   **b.**

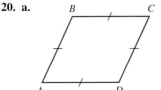   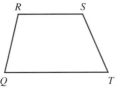

21. a. 12   **b.** 13   **c.** 90°   **d.** 5   **22. a.** 10   **b.** 65°   **c.** 115°
d. 115°   **23.** 1,440°   **24. a.** 6   **b.** 60°   **25.** 188 in.
26. 15.2 m   **27.** 13 ft, 4 ft   **28.** 360 cm^2
29. $(b^2 - 2b)$ square units   **30.**

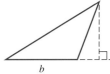

31. $800   **32.** 144   **33.** No; The area is 25 ft^2.   **34.** 120 in.^{2}
35. a. $\overline{CD}, \overline{AB}$ **b.** $\overline{AB}$ **c.** $\overline{OA}, \overline{OC}, \overline{OD}, \overline{OB}$ **d.** $\overparen{BD}$,
$\overparen{DAB}$, or $\overparen{DCB}$ **e.** semicircle   **36.** π **37.** 21π cm ≈ 66.0 cm
38. $(20 + 8\pi)$ cm ≈ 45.1 cm   **39.** $\frac{53.4}{\pi}$ in. ≈ 17.0 in.
40. 225π m$^2 \approx 706.9$ m^2 **41.** $(20.25 - 5.0625\pi)$ mi$^2 \approx 4.3$ mi^2
42. $112   **43. a.** $\sqrt{\frac{38.5}{\pi}}$ ft ≈ 3.5 ft   **b.** 7.0 ft   **c.** 22.0 ft
44. a. 47°   **b.** 313°   **c.** 33°   **45.** $\frac{11\pi}{5}$ in. ≈ 6.9 in.
46. $\frac{20{,}000\pi}{3}$ ft$^2 \approx 20{,}944.0$ ft^2

Study Set   Section 8   (page 101)

1. Congruent   **3.** congruent   **5.** similar   **7. a.** No. They have
different sizes.   **b.** Yes. They have the same shape.
9. PRQ **11.** MNO **13.** $\angle A \cong \angle B$, $\angle Y \cong \angle T$, $\angle Z \cong \angle R$, $\overline{YZ} \cong$
$\overline{TR}$, $\overline{AZ} \cong \overline{BR}$, $\overline{AY} \cong \overline{BT}$ **15.** congruent   **17.** angle, angle
19. 100   **21.** 5.4   **23.** proportional   **25.** congruent
27. is congruent to
29.

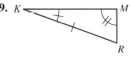   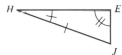

31. $\overline{DF}, \overline{AB}, \overline{EF}, \angle D, \angle B, \angle C$ **33. a.** $\angle B \cong \angle M, \angle C \cong \angle N$,
$\angle D \cong \angle O, \overline{BC} \cong \overline{MN}, \overline{CD} \cong \overline{NO}, \overline{BD} \cong \overline{MO}$ **b.** 72°   **c.** 10 ft
d. 11 ft   **35.** yes, SSS   **37.** not necessarily   **39. a.** $\angle L \cong \angle H$,
$\angle M \cong \angle J, \angle R \cong \angle E$ **b.** MR, LR, LM **41.** yes
43. not necessarily   **45.** yes   **47.** not necessarily   **49.** yes
51. not necessarily   **53.** 8, 35   **55.** 60, 38   **57.** true
59. false. The angles must be between the congruent
sides.   **61.** yes, SSS   **63.** yes, SAS   **65.** yes, ASA
67. not necessarily   **69.** 80°; 2 yd   **71.** 19°; 14 m
73. 6 mm   **75.** 50°   **77.** $\frac{25}{6} = 4\frac{1}{6}$ **79.** 16   **81.** 17.5 cm
83. 36 ft   **85.** 59.2 ft

Study Set   Section 9   (page 117)

1. hypotenuse, legs   **3.** Pythagorean   **5.** isosceles   **7.** a^2, b^2,
c^2 **9.** 30°, 60°   **11.** 6, -6, -6
13. a.   **b.**   **c.**

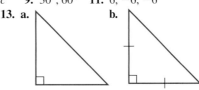   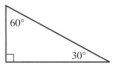

15. Subtract 25 from both sides.   **17. a.** 9   **b.** $3\sqrt{3}$ **c.** 3
d. $10\sqrt{2}$ **19.** 64, $\sqrt{100}$, 10   **21.** $c = a\sqrt{2}$ **23.** 10 ft
25. 13 m   **27.** 73 mi   **29.** 137 cm   **31.** 24 cm   **33.** 80 m
35. 20 m   **37.** 19 m   **39.** $\sqrt{24}$ cm $= 2\sqrt{6}$ cm ≈ 4.9 cm
41. $\sqrt{208}$ m $= 4\sqrt{13}$ m ≈ 14.4 m   **43.** $\sqrt{90}$ in. $= 3\sqrt{10}$ in. $\approx$
9.5 in.   **45.** $\sqrt{20}$ in. $= 2\sqrt{5}$ in. ≈ 4.5 in.   **47.** 8; 6, 10
49. 3, 4, 5   **51.** no   **53.** yes   **55.** 2 cm, $2\sqrt{2}$ cm $\approx$
2.83 cm   **57.** $3.2\sqrt{2}$ ft ≈ 4.53 ft   **59.** $\frac{3\sqrt{2}}{2}$ yd ≈ 2.12 yd
61. $5\sqrt{2}$ in. ≈ 7.07 in.   **63.** 10 mm; $5\sqrt{3}$ mm ≈ 8.66 mm

65. $75\sqrt{3}$ cm $\approx$ 129.90 cm; 150 cm **67.** leg: $\frac{40\sqrt{3}}{3}$ cm $\approx$ 23.09 cm;
hypotenuse: $\frac{80\sqrt{3}}{3}$ cm $\approx$ 46.19 cm **69.** leg: $\frac{55\sqrt{3}}{3}$ mm $\approx$ 31.75 mm;
hypotenuse: $\frac{110\sqrt{3}}{3}$ mm $\approx$ 63.51 mm **71.** shorter leg: 50 ft;
longer leg: $50\sqrt{3}$ ft $\approx$ 86.60 ft **73.** shorter leg: 0.75 ft; longer leg:
$0.75\sqrt{3}$ ft $\approx$ 1.30 ft **75.** $\sqrt{51}$ in. $\approx$ 7.1 in. **77.** $\sqrt{23}$ ft $\approx$ 4.8 ft
79. $\sqrt{98}$ cm $= 7\sqrt{2}$ cm **81.** 8 ft, 15 ft, 17 ft **83.** $(5\sqrt{2}, 0)$,
$(0, 5\sqrt{2})$, $(-5\sqrt{2}, 0)$, $(0, -5\sqrt{2})$; (7.07, 0), (0, 7.07), (-7.07, 0),
(0, -7.07) **85.** $10\sqrt{3}$ mm $\approx$ 17.32 mm **87.** $\sqrt{18,100}$ ft $=$
$10\sqrt{181}$ ft $\approx$ 134.54 ft **89.** about 0.13 ft **91.** 5 m, 12 m, 13 m

Study Set   Section 10   (page 131)

1. space **3.** right, oblique **5.** cube **7.** perpendicular
9. cubic **11. a.** circular cone **b.** sphere **c.** right circular cylinder **d.** right triangular prism **e.** square pyramid **f.** right
regular pentagonal prism **13.**

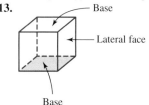

15.   **17.**

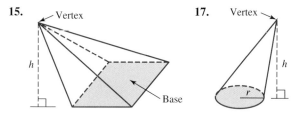

19. mi^3, cubic inches, m^3 **21. a.** perimeter **b.** volume
c. area **d.** volume **e.** area **f.** circumference
23. a. 50π **b.** $\frac{500}{3}\pi$ **25. a.** cubic inch **b.** 1 cm^3
27. a right angle **29. a.** 2, hexagon **b.** 6, rectangles
c. right hexagonal prism **31. a.** 2, triangle
b. 3, rectangles **c.** oblique triangular prism **33. a.** 1, pentagon **b.** 5, triangles **c.** right pentagonal pyramid **35. a.** 1,
rectangle **b.** 4, triangles **c.** oblique rectangular pyramid
37. 27 **39.** 1,000,000,000 **41.** 56 ft^3
43. 125 in.^{3} **45.** 120 cm^3 **47.** 1,296 in.^{3}
49. 700 yd^3 **51.** 32 ft^3 **53.** 69.72 ft^3 **55.** 6 yd^3
57. 27 in. **59.** 57 cm **61.** 192π ft^3 $\approx$ 603.19 ft^3
63. $3,150\pi$ cm^3 $\approx$ 9,896.02 cm^3 **65.** 39π m^3 $\approx$ 122.52 m^3
67. 189π yd^3 $\approx$ 593.76 yd^3 **69.** 288π in.^{3} $\approx$ 904.8 in.^{3}
71. $\frac{32}{3}\pi$ cm^3 $\approx$ 33.5 cm^3 **73.** 4 cm **75.** 18 ft
77. 486π in.^{3} $\approx$ 1,526.81 in.^{3} **79.** 432π m^3 $\approx$ 1,357.17 m^3
81. 60 cm^3 **83.** 100π cm^3 $\approx$ 314.16 cm^3 **85.** 400 m^3
87. 48 m^3 **89.** 3 ft **91.** 8 yd **93.** 15 ft **95.** 30 ft^3
97. 0.416π m^3 $\approx$ 1.31 m^3 **99.** $4,500\sqrt{3}$ ft^3 $\approx$ 7.794.23 ft^3
101. 576 cm^3 **103.** 180π cm^3 $\approx$ 565.49 cm^3
105. $\frac{1}{8}$ in.^{3} = 0.125 in.^{3} **107.** 2.125 **109.** 63π ft^3 $\approx$ 197.92 ft^3
111. $\frac{32,000}{3}\pi$ ft^3 $\approx$ 33,510.32 ft^3 **113.** 8:1 **115. a.** $2,250\pi$ in.^{3} $\approx$
7,068.58 in.^{3} **b.** 30.6 gal

Study Set   Section 11   (page 145)

1. surface area **3.** lateral **5.** bases **7.** right **9.** ft^2, cm^2,
square yards **11. a.** bases **b.** lateral **13.** 180 in.

15.

Figure	Total Surface Area Formula
Right prism	$TSA = 2B + hp$
Right circular cylinder	$TSA = 2\pi r^2 + 2\pi rh$
Right regular pyramid	$TSA = B + \frac{1}{2}ps$
Right circular cone	$TSA = \pi r^2 + \pi rs$
Sphere	$SA = 4\pi r^2$

17. a. $11\sqrt{3}$ in. **b.** $121\sqrt{3}$ in.^{2} $\approx$ 209.58 in.^{2}
19. a. 41π **b.** 44 **21. a.** the total surface area of a right
prism **b.** the area of the base of the prism **c.** the height of the
prism **d.** the perimeter of the base of the prism **23.** 5,040 in.^{2}
25. 600 in.^{2} **27.** 72 in.^{2} **29.** 9,280 in.^{2} **31.** 309 cm^2
33. 374 in.^{2} **35.** 130π ft^2 $\approx$ 408.4 ft^2 **37.** 315π ft^2 $\approx$ 989.6 ft^2
39. $9\sqrt{3}$ m^2; $(9\sqrt{3} + 81)$ m^2 $\approx$ 96.6 m^2 **41.** $225\sqrt{3}$ m^2;
$(225\sqrt{3} + 1,800)$ m^2 $\approx$ 2,189.7 m^2 **43.** 224π in.^{2} $\approx$ 703.7 in.^{2}
45. 980π in.^{2} $\approx$ 3,078.8 in.^{2} **47.** 36π in.^{2} $\approx$ 113.1 in.^{2}
49. 225π in.^{2} $\approx$ 706.9 in.^{2} **51.** 604 ft^2 **53.** $(12\sqrt{3} + 72)$ ft^2 $\approx$
92.8 ft^2 **55.** $4\sqrt{3}$ cm^2; $(8\sqrt{3} + 60)$ cm^2 $\approx$ 73.9 cm^2
57. 380 yd^2 **59.** 800 yd^2 **61.** $150\sqrt{3}$ m^2; $(150\sqrt{3} + 630)$ m^2 $\approx$
889.8 m^2 **63.** 9 ft **65.** 12 m **67.** 3 ft
69. 175π in.^{2} $\approx$ 549.8 in.^{2} **71.** $(432\sqrt{3} + 720)$ mm^2 $\approx$
1,468.2 mm^2 **73.** 4,080 mm^2 **75.** 93,600 ft^2
77. 4.44π in.^{2} $\approx$ 14 in.^{2}

Cumulative Review Problems   Sections 8–11
(page 150)

1. h **2.** e **3.** g **4.** a **5.** c **6.** i **7.** b **8.** j **9.** f **10.** d
11. $\angle D$, $\angle E$, $\angle F$, $\overline{DF}$, $\overline{DE}$, $\overline{EF}$ **12. a.** congruent, SSS
b. congruent, ASA **c.** not necessarily congruent
d. congruent, SAS **13. a.** 8 in. **b.** 50° **14. a.** yes **b.** yes
15. a. 6 m **b.** 12 m **16.** 21 ft **17. a.** 26 cm **b.** $\sqrt{28}$ in. $=$
$2\sqrt{7}$ in. $\approx$ 5.3 in. **18. a.** hypotenuse: $2\sqrt{2}$ ft $\approx$ 2.83 ft; other leg:
2 ft **b.** shorter leg: $\frac{7\sqrt{3}}{3}$ yd $\approx$ 4.04 yd; hypotenuse: $\frac{14\sqrt{3}}{3}$ yd $\approx$
8.08 yd **c.** longer leg: $9\sqrt{3}$ cm $\approx$ 15.59 cm; hypotenuse: 18 cm
19. 31.4 in. **20.** 3 ft, 4 ft, 5 ft **21.** 1,728 in.^{3} **22. a.** 125 cm^3
b. 480 m^3 **c.** 250π in.^{3} $\approx$ 785.40 in.^{3} **d.** 600 in.^{3}
e. $9,020,833\frac{1}{3}$ ft^3 **23.** $11,250\pi$ ft^3 $\approx$ 35,343 ft^3 **24.** 4 in.
25. a. 61.78 ft^2 **b.** 60π yd^2 $\approx$ 188.50 yd^2 **c.** 360 mi^2
d. 100π in.^{2} $\approx$ 314.16 in.^{2} **26.** 24 ft

Study Set   Appendix I   (page A-5)

1. Inductive **3.** circular **5.** alternating **7.** alternating
9. 10 A.M. **11.** 17 **13.** 27 **15.** 3 **17.** -17 **19.** R **21.** e
23.   **25.**   **27.**

29.   **31.** D **33.** Maria **35.** 6

37. 9 **39.** I **41.** W **43.**   **45.** K **47.** 6

49. 3 **51.** -11 **53.** 9 **55.** cage 3 **57.** B, D, A, C
59. 18,935 **61.** 0